TEXTBOOK OF ANIMAL GENETICS AND BREEDING

Edited by
Nada Ben Abdallah

ArclerPress

Textbook of Animal Genetics and Breeding

Edited by Nada Ben Abdallah

Edition 2018

ArclerPress

Published by
Arcler Press LLC
708 3rd Avenue, 6th Floor
New York
NY 10017
United States of America
www.arclerpress.com

ISBN: 978-1-68094-591-1 (Hardback)

Printed in United States of America

Library of Congress Control Number: 2017933188

About the Editor

Nada Ben Abdallah

Nada graduated from the University of Zurich in 2010 and holds a PhD in neuroscience. Her research focuses on elucidating the processes of learning and memory as well as the affective processes and their underlying neural mechanisms. Her work so far involved normal and diseased animal models with a main interest in neurodegenerative diseases. She is currently working as a freelance researcher and science editor.

Table of Contents

3 METHODS OF ANIMAL BREEDING

4 ARTIFICIAL INSEMINATION IN ANIMALS

5 ANIMAL SEXUAL BEHAVIOR

6 BREED STANDARDS

7 SYSTEMS OF MATING

8 GENOTYPES: GENETIC MAKEUPS OF ORGANISMS

9 CUNICULTURE

List of Abbreviations

ASO	Allele specific Oligonucleotide
ASPCA	American Society for the Prevention of Cruelty to Animals
AFLP	Amplified Fragment Length Polymorphism
AI	Artificial Insemination
ARTs	Assisted Reproductive Technology
IPCBA	International Progressive Cat Breeders Alliance
ICI	Intra-Cervical Insemination
IVI	Intravaginal Insemination
CFA	Cat Fanciers' Association
DIPI	Direct Intraperitoneal Insemination
MF	Magnetic Field-Free Space
MUP	Major Urinary Protein
MSI	Microsatellite Instability
t-RFLP	Terminal Restriction Fragment Length Polymorphism
TICA	The International Cat Association
VI	Vaginal Insemination

Preface

Animal breeding and genetics is the branch of science concerned with maximizing desirable genetic traits in domesticated animals. The study of animal genetics involves topics such as gene expression, color genetics and animal breeding. Animal genetics focuses on various aspects of animal heredity, or the passing of traits from one generation to the next. The field encompasses topics such as genetic variability, genetic testing, and animal breeding. Animal geneticists have identified elements within genes that can enhance animal growth, health, and ability to utilize nutrients. These genetic advances can increase production while reducing environmental impacts. Animals and livestock contribute 40 percent of the global value of agricultural output and contribute to the livelihoods and food security of almost a billion people worldwide. Advances in animal breeding, genetics, and genomics are facilitating a more efficient industry. For example, the number of cattle has decreased over the past decade, yet the total production of beef and milk has increased. This was largely possible because genetic advancements led better animal feed efficiency, which is critical to improving livestock production and lowering costs for producers. When breeding is between animals of the same breed for 4-6 generations, it is called inbreeding. Inbreeding may be explained by taking an example of cows and bulls. Superior cows and superior bulls of the same breed are identified and mated. The progeny obtained from such mating are evaluated and superior males and females are identified for further mating. A superior female, in the case of cattle, is the cow that produces more milk per lactation. Thus inbreeding is necessary to develop a pure line in any animal. Inbreeding exposes harmful recessive genes that are eliminated by selection. It also helps in accumulation of superior genes and elimination of less desirable genes. Out breeding is the breeding between the unrelated animals which may be between individuals of the same breed (but having no common ancestors) or between different breeds (cross breeding) or different species (interspecific hybridization).

Content Coverage

Chapters One and *Two* present an introduction of animal breeding and animal breeds, correspondingly. Breeds are formed through genetic isolation and either natural adaptation to the environment or selective breeding, or a combination of the two.

Chapters Three and *Four* present the methods of animal breeding and artificial insemination in animals, respectively. In animal breeding, crossbreed describes crosses within a single species, while hybrid refers to crosses between different species.

Chapter Five is intended to focus on animal sexual behavior. Animal sexual Behavior takes many different forms, including within the same species. Common mating or reproductively motivated systems include monogamy, polyandry, polygamy, and promiscuity.

Chapters Six and *Seven* present an overview of breed standards and systems of mating, respectively. Breed standards are designed to protect the integrity of purebred animals and their heritage by ensuring that only the finest representatives of the breed are identified as members of the breed. For animals, mating strategies include random mating, disassortative mating, assortative mating, or a mating pool.

Chapter Eight presents a focus on genotypes. The genotype is the part (DNA sequence) of the genetic makeup of a cell, and therefore of an organism or individual, which determines a specific characteristic (phenotype) of that cell/organism/individual.

Chapter Nine focus on cuniculture is the agricultural practice of breeding and raising domestic rabbits, usually for their meat, fur, or wool.

CHAPTER 1

Introduction to Animal Breeding

Chapter Contents

- **Introduction**
- **Breeding Stock**
- **Backyard Breeder**

Animal breeding, controlled propagation of domestic animals in order to improve desirable qualities. Humanity has been modifying domesticated animals to better suit human needs for centuries. Selective breeding involves using knowledge from several branches of science. These include genetics, statistics, reproductive physiology, computer science, and molecular genetics. This article discusses the basic principles of how populations of animals can be changed by application of these principles, and a brief discussion of molecular genetics, immunogenetics, and newer reproductive technologies is included. The fundamental biological principles underlying animal breeding are discussed in the articles heredity and animal reproductive system.

INTRODUCTION

Animal breeding is a branch of animal science that addresses the evaluation (using best linear unbiased prediction and other methods) of the genetic value (estimated breeding value, EBV) of livestock. Selecting for breeding animals with superior EBV in growth rate, egg, meat, milk, or wool production, or with other desirable traits has revolutionized livestock production throughout the world. The scientific theory of animal breeding incorporates population genetics, quantitative genetics, statistics, and recently molecular genomics and is based on the pioneering work of Sewall Wright, Jay Lush, and Charles Henderson.

BREEDING STOCK

Breeding stock is a group of animals used for the purpose of planned breeding. When individuals are looking to breed animals, they look for certain valuable traits in purebred animals, or may intend to use some type of crossbreeding to produce a new type of stock with different, and presumably superior abilities in a given area of endeavor. For example, when breeding swine the "breeding stock should be sound, fast growing, muscular, lean, and reproductively efficient." The "subjective selection of breeding stock" in horses has led to many horse breeds with particular performance traits.

Purebred Breeding

Mating animals of the same breed for maintaining such breed is referred to as purebred breeding. Opposite to the practice of mating animals of different breeds, purebred breeding aims to establish and maintain stable traits that animals will pass to the next generation. By "breeding the best to the best," employing a certain degree of inbreeding, considerable culling, and selection for "superior" qualities, one could develop a bloodline or "breed" superior in certain respects to the original base stock.

Such animals can be recorded with a breed registry, the organisation that maintains pedigrees and/or stud books. The observable phenomenon of hybrid vigor stands in contrast to the notion of breed purity.

Backyard Breeding

In the United States, a backyard breeder is someone who breeds animals, often without registration and with a focus on profit. In some cases the animals are inbred narrowly for looks with little regard to health. The term is considered derogatory. If a backyard dog breeder has a significant number of breeding animals, they become associated with puppy mills. Most puppy mills are licensed with the USDA.

BACKYARD BREEDER

Backyard breeder is a term used to describe amateur animal breeders whose breeding is considered substandard, with little or misguided effort towards ethical, selective breeding. Unlike puppy mills and other animal mill operations, backyard breeders breed on a small scale, usually at home with their own pets (hence the "backyard" description), and may be motivated by things other than monetary profit, such as curiosity, to gain new pets, or to show children "the miracle of birth".

The term most often is used to describe certain types of breeders of dogs and horses. The term is used in this sense by the Animal Welfare community, The American Society for the Prevention of Cruelty to Animals (ASPCA), larger established breeders and breed clubs in contrast to the more positive term, «reputable breeder» that describes operations that use responsible methodology and practices.

It implies either or both of home breeding for non-commercial reasons or a for-profit small commercial operation that does not adhere to good breeding, care and sale practices. Larger commercial operations of a similar type that breed dogs are usually termed a *puppy mill* (especially in North America) or *puppy farm*. However, as large kennels usually require licensing, many puppy mills are licensed with the USDA.

Many small breeders resent the term and its use. Some do so because they assert that they run small but high-quality "boutique" operations. Others argue that their unregistered animals are highly desirable as companion animals and economical for ordinary people to purchase. Both groups deny that they contribute to low quality within the registered breed or produce unwanted animals.

Undesirable Characteristics

This type of breeder is usually regarded by the Animal Welfare, ASPCA, and other groups as likely to exhibit one or more of the following characteristics:

- Ignorance of selective breeding goals and techniques, and lack of familiarity with the breed standard of the type of animal being bred.
- Exclusive focus on the breed standard involving little genetic screening or co-efficient of Inbreeding calculations.
- Breeding of a working breed for appearance rather than working ability. This is a criticism also levelled at 'reputable' breeders who breed for the show ring - in some cases distinct **working** and **show** strains have emerged.
- Lack of adequate veterinary care and maintenance.
- Excessive breeding from individual females, to the detriment of their health.
- Sale of animals with genetic disorders or undisclosed illnesses before they become evident to buyers.
- Lack of screening of potential owners or the provision of suitable information to prevent buyers from purchasing an animal that may be inappropriate for them or their lifestyle.
- Breeding of animals for illegitimate reasons, such as for horse slaughter or the usually illegal sports of baiting and dog fighting or to defend venues of criminal activity. Dog fanciers generally believe that such ill-bred dogs are the reason for the bad reputation of some breeds in the public perception, and the resulting breed-specific legislation. The production of «PMU foals» from pregnant mares bred solely for their urine production (used in the making of the drug Premarin) is also widely condemned due to the frequency with which the progeny of such breedings are shipped directly to slaughter.
- Breeding without concern for the possibility of finding homes, though commercial means or otherwise, for offspring, thus adding to the population of unwanted dogs and to the slaughter of horses for meat.

With horses, the USA horse slaughter issue is closely linked to backyard breeding, as many poorly bred, unregistered horses have no value other than to horsemeat buyers.

Opposition

There are campaigns to reduce or eliminate this type of breeding in several countries as well as video campaigns by groups like ASPCA that emphasize the importance of understanding the responsibilities associated with owning an animal. The primary concern of the Animal Welfare community is that this type of breeding is the major source of pet animals and that overproduction has led to overpopulation and hence to welfare issues and the putting down of animals for no other reason than the lack of suitable home. The glamorization of Chihuahuas, which some claim has been generated by Hollywood, has had devastating outcomes for dogs and their adopted owners. Specifically, dog owners like Paris Hilton and commercial-star Gidget the Taco Bell chihuahua as well as films like Beverly Hills Chihuahua have been directly cited as blameworthy. Most dog breed societies and Kennel Clubs have similar concerns but place rather more stress on issues relating to breed standardization and quality.

There are many animal adoption alternatives that potential pet owners can consider, including animal shelters and the umbrella organization The Shelter Pet Project, animal rescue organizations, online organizations such as Pet Finder that directly connect prospective adopters with available animals, and print media such as local newspapers).

REFERENCES

1. Oklahoma Cooperative Extension Service, Division of Agricultural Sciences and Natural Resources (n.d.). "Selection of Swine Breeding Stock" (PDF). pp. 1–4.
2. James Warren Evans (1992). Horse breeding and management. Elsevier Health Science. ISBN 978-0-444-88282-0.
3. The Obama family dog saga. LA Times.
4. Rescue groups paint a sad story of Iowa's puppy mills. The Messenger.
5. Rescue groups paint a sad story of Iowa's puppy mills. The Messenger.
6. ASPCA Cruelty Glossary ASPCA website/
7. GREYHOUND OVER BREEDING Campaigning for the Abolition of Betting on Greyhounds
8. Puppies are not Toys ASPCA YouTube
9. California animal shelters' plague: Not a rain of frogs, but a rain of Chihuahuas LA Times
10. 2-year-old Chihuahua Rosie Before She was Rescued Everything Rosie
11. Taco Bell Dog Spurs Pet Revolution Philly.com
12. The Chihuahua Boom About.com
13. California's chihuahua problem? Blame it on Paris Hilton The Guardian
14. Shelters

15. A directory of European animal shelters is available at http://www.bringfido.com/resource/animal_shelters/region/europe/.
16. Animal Shelters and Rescues (via Petfinder.com)
17. For example, the Chicago Tribune runs a "Pets In Need Of Homes" feature (see http://www.chicagotribune.com/features/chi-ugc-animal-rescue-dog-cats-adoption,0,5554945.photogallery)

CHAPTER 2

Animal Breeds

Chapter Contents

- **Introduction**
- **Breeds**
- **List of Cattle Breeds**
- **List of Chicken Breeds**
- **List by Place of Origin**
- **List of Donkey Breeds**

INTRODUCTION

The following list of breeds includes only domestic cat breeds and domestic × wild hybrids. The list includes established breeds recognized by various cat registries, new and experimental breeds, distinct domestic populations not being actively developed, and lapsed breeds. As of 2016, the International Progressive Cat Breeders Alliance (IPCBA) recognizes 73 cat breeds while The International Cat Association (TICA) recognizes 58, the Cat Fanciers` Association (CFA) recognizes 44, and Fédération Internationale Féline (FIFe) recognizes 43.

Inconsistency in breed classification among registries means that an individual animal may be considered different breeds by different registries. For example, TICA`s Himalayan is considered a colorpoint version of the Persian by CFA, while CFA`s Javanese is considered a color variation of the Balinese by TICA.

The domestic short-haired and domestic long-haired cat types are not breeds but terms used (with various spellings) in the cat fancy to describe cats of a general type, by coat length, that do not belong to a particular breed.

BREEDS

Breed	Country	Origin	Body type	Coat	Pattern	Image
Abyssinian	Ethiopia	Natural/Standard		Short	Ticked	
Aegean	Greece	Natural/Standard		Semi-long	Bi- or tri-colored	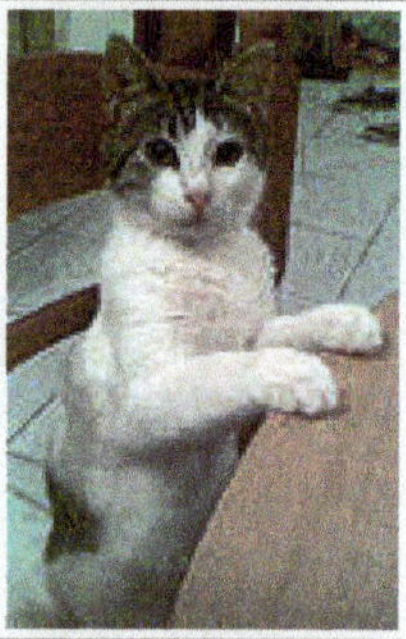
American Curl	United States	Mutation		Short/ Long	All	
American Bobtail	United States	Mutation		Short/ Long	All	

Breed	Country	Origin	Body type	Coat	Pattern	Image
American Shorthair	United States	Natural		Short	All but colorpoint	
American Wirehair	United States	Mutation		Rex	All but colorpoint	
Arabian Mau	Arabian Peninsula	Natural		Short		
Australian Mist	Australia	Crossbreed	Moderate	Short	Spotted and Classic tabby	
Asian	developed in the United Kingdom (founding stock from Asia)			Short	Evenly solid	
Asian Semi-longhair	United Kingdom	Crossbreed		Semi-long	Solid	
Balinese	developed in the United States (founding stock from Thailand)	Crossbreed	Oriental	Long	Colorpoint	

Breed	Country	Origin	Body type	Coat	Pattern	Image
Bambino	United States	Crossbreed		Hairless/ Furry down		
Bengal	developed in the United States (founding stock from Asia)	Hybrid		Short	Spotted/ Marbled	
Birman	Burma	Natural		Semi Long	Colorpoint	
Bombay	developed in the United States (founding stock from Asia)	Crossbred	Moderate	Short	Solid	
Brazilian Shorthair	Brazil	Natural		Short	All	
British Semi-longhair	United Kingdom		Cobby	Medium	All	
British Shorthair	United Kingdom	Natural	Cobby	Short	All	

Breed	Country	Origin	Body type	Coat	Pattern	Image
British Longhair	United Kingdom		Cobby	Long		
Burmese	Burma and Thailand	Natural		Short	Solid	
Burmilla	United Kingdom	Crossbreed		Short/ Long		
California Spangled	United States	Crossbreed		Short	Spotted	
Chantilly-Tiffany	United States					
Chartreux	France	Natural	Cobby	Short	Solid	
Chausie	France	Hybrid		Short	Ticked	
Cheetoh	United States	Hybrid Crossbreed		Short	Spotted	

Breed	Country	Origin	Body type	Coat	Pattern	Image
Colorpoint Shorthair				Short		
Cornish Rex	United Kingdom (England)	Mutation		Rex	All	
Cymric, or Manx Longhair	United Kingdom (Isle of Man)	Natural/Mutation		Long		
Cyprus	Cyprus	Natural	Lean and muscular	All	All	
Devon Rex	United Kingdom (England)	Mutation	Oriental	Rex	All	
Donskoy, or Don Sphynx	Russia			Hairless		
Dragon Li	China	Natural		Short	Striped tabby	
Dwarf cat, or Dwelf		Crossbreed			Hairless	
Egyptian Mau	Egypt	Natural		Short	Spotted	

Breed	Country	Origin	Body type	Coat	Pattern	Image
European Shorthair	Finland and Sweden	Natural		Short		
Exotic Shorthair	United States	Crossbreed	Cobby	Short	All	
Foldex (Official Breed - CCA/AFC)	Canada	Crossbreed	Cobby	Short	All	
German Rex	East Germany	Mutation		Rex		
Havana Brown	United Kingdom			Short	Solid	
Highlander	United States	Crossbreed	Moderate	Short/ Long	All	
Himalayan, or Colorpoint Persian	Nepal	Crossbreed	Cobby	Long	Colorpoint	
Japanese Bobtail	Japan	Natural	Moderate	Short/ Long	All but colorpoint and ticked	

Breed	Country	Origin	Body type	Coat	Pattern	Image
Javanese	United States	Crossbreed	Oriental	Long/ Short	Colorpoint	
Kurilian Bobtail	Western Russia	Natural				
Khao Manee	Thailand	Natural		Short	Solid	
Korat	Thailand	Natural		Short	Solid	
Korean Bobtail	Korea	Natural		Short/ Long	Colorprint	
Korn Ja	Thailand	Natural		Short	Solid	
Kurilian Bobtail, or Kuril Islands Bobtail	Eastern Russia, Japan	Natural	Semi-Cobby	Short/ Long		
LaPerm	United States	Mutation	Moderate	Rex	All	

Breed	Country	Origin	Body type	Coat	Pattern	Image
Lykoi	United States	Natural/Mutation		Partly Hairless	Ticked	
Maine Coon	United States	Natural	Large	Long	All but colorpoint and ticked	
Manx	United Kingdom (Isle of Man)	Natural/Mutation		Short/ Long	All but colorpoint	
Mekong Bobtail	Russia	Natural/Mutation		Short	Colorpoint	
Minskin	United States	Crossbreed	Semi-Cobby	Short/ Hairless	All	
Munchkin	United States	Mutation				
Nebelung	United States			Semi-long	Solid	
Napoleon				Long/ short	Varied	

Breed	Country	Origin	Body type	Coat	Pattern	Image
Norwegian Forest cat	Norway	Natural		Long	All but colorpoint	
Ocicat	United States	Crossbreed		Short	Spotted	
Ojos Azules	United States					
Oregon Rex	United States	Mutation		Rex		
Oriental Bicolor			Oriental		Bicolor	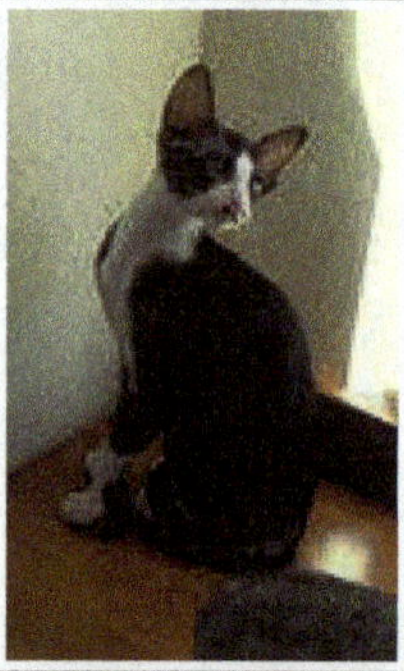
Oriental Shorthair			Oriental	Short	All but colorpoint	
Oriental Longhair			Oriental	Semi-long		
PerFold Cat (Experimental Breed - WCF)	Europe	Crossbreed	Cobby	Long	All	

Breed	Country	Origin	Body type	Coat	Pattern	Image
Persian (Modern Persian Cat)	Iran (Persia)	Crossbreed	Cobby	Long	All	
Persian (Traditional Persian Cat)	Greater Iran	Natural	Cobby	Long	All	
Peterbald	Russia	Crossbreed	Stocky	Hairless	All	
Pixie-bob	United States	Natural		Short	Spotted	
Raas	Indonesia	Natural		Short		
Ragamuffin	United States	Crossbreed	Cobby	Long	All	
Ragdoll	United States	Crossbreed	Cobby	Long	Colorpoint/ Mitted/ Bicolor	

Breed	Country	Origin	Body type	Coat	Pattern	Image
Russian Blue	Russia	Natural		Short	Solid	
Russian White, Black and Tabby	Australia	Crossbreed		Short		
Sam Sawet	Thailand	Natural		Short	Solid	
Savannah	United States	Hybrid		Short	Spotted	
Scottish Fold	United Kingdom (Scotland)	Natural/Mutation	Cobby	Short/ Long	All	
Selkirk Rex	United States	Mutation/Cross		Rex (Short/ Long)	All	
Serengeti	United States	Hybrid Crossbreed		Short	Spotted	
Serrade petit	France	Natural		Short		

Breed	Country	Origin	Body type	Coat	Pattern	Image
Siamese	Thailand	Natural	Oriental	Short	Colorpoint	
Siberian	Russia	Natural	Semi-Cobby	Semi-long	All	
Singapura	Singapore	Natural		Short	Ticked	
Snowshoe	United States	Crossbreed		Short	Colorpoint	
Sokoke	Kenya	Natural		Short	Classic tabby with ticking	
Somali	Somalia	Mutation		Long	Ticked	
Sphynx	Canada	Mutation	Stocky	Hairless	All	

Breed	Country	Origin	Body type	Coat	Pattern	Image
Suphalak	Thailand	Natural		Short	Solid	
Thai	Thailand	Natural		Short	Colorpoint	
Thai Lilac	Thailand	Natural		Short	Solid	
Tonkinese	Canada	Crossbreed		Short	Colorpoint/ Mink/Solid	
Toyger	United States	Crossbreed	Moderate	Short	Mackerel	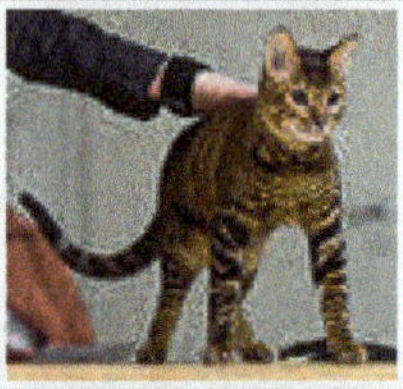
Turkish Angora	Turkey	Natural		Semi-long	All but colorpoint	
Turkish Van	developed in the United Kingdom (founding stock from Turkey)	Natural		Semi-long	Van	

Breed	Country	Origin	Body type	Coat	Pattern	Image
Ukrainian Levkoy	Ukraine			Hairless		

LIST OF CATTLE BREEDS

Breeds fall into two main types, regarded as either two closely related species, or two subspecies of one species. *Bos indicus* (or *Bos taurus indicus*) cattle, also called zebu, are adapted to hot climates. *Bos taurus* (or *Bos taurus taurus*) are the typical cattle of Europe, north-eastern Asia, and parts of Africa – they are referred to in this list as "taurine" cattle, and many are adapted to cooler climates. *Taurus/indicus* hybrids are widely bred in many warmer regions, combining characteristics of both the ancestral types (such as the Sanga cattle of Africa).

In some parts of the world further species of cattle are found (both as wild and domesticated animals), and some of these are related so closely to taurine and indicus cattle that interspecies hybrids have been bred. Examples include the Dwarf Lulu cattle of the mountains of Nepal with yak blood, the Beefalo of North America with bison genes, the Selembu breed of India and Bhutan with gayal genes. The Madura breed of Indonesia may have banteng in its parentage. In addition to these fertile hybrids, there are infertile crossings such as the male Dzo of Nepal, a cattle-yak crossing which is bred for agricultural work - like the mule they have to be continually bred from the parent species.

A

Alphabetical List - A							
Breed	**Image**	**Subspecies**	**Country/ region of origin**	**Meat**	**Dairy**	**Draught**	**Other**
Abigar		Taurus	Ethiopia	—	Dairy	—	—
Abondance (cattle)		Taurus	France	Meat	Dairy	—	—
Abyssinian Shorthorned Zebu		Indicus	Ethiopia	—	—	Draught	—
Aceh		—	Indonesia	—	—	—	—

Alphabetical List - A							
Breed	**Image**	**Subspecies**	**Country/ region of origin**	**Meat**	**Dairy**	**Draught**	**Other**
Achham		Indicus	Nepal	—	Dairy	—	—
Adamawa (cattle)		Taurus	Nigeria	Meat	Dairy	Draught	—
Adaptaur		Taurus/ Indicus hybrid	Australia	Meat	—	—	—
Africangus		Taurus/ Indicus hybrid	United States	Meat	—	—	—
Afrikaner		Indicus	South Africa	Meat	—	—	—
Agerolese		Taurus	Italy	—	Dairy	—	—
Akaushi		—	Japan	Meat	—	—	one of the four Kobe breeds
Ala Tau		Taurus	Kazakhstan	Meat	Dairy	—	—
Alambadi		Indicus	India	—	—	—	—
Alatau cattle		Taurus	Kazakhstan	Meat	Dairy	—	—
Albanian (cattle)		Taurus	Albania	—	Dairy	Draught	—
Albera		Taurus	Catalonia	Meat	—	—	Semi-feral
Alderney		Taurus	Channel Islands	—	Dairy	—	—
Alentejana		Taurus	Portugal	Meat	—	Draught	—
Aleutian wild cattle		—	United States	—	—	—	Feral
Aliab Dinka		—	South Sudan	—	—	—	—
Alistana-Sanabresa		—	Spain	—	—	—	—
Allmogekor		Taurus	Sweden	Meat	—	Draught	—
Alur		—	Democratic Republic of the Congo	—	—	—	—

Alphabetical List - A							
Breed	**Image**	**Subspecies**	**Country/ region of origin**	**Meat**	**Dairy**	**Draught**	**Other**
American		Taurus/ Indicus hybrid	United States	Meat	—	—	—
American Angus		—	United States	—	—	—	—
American Beef Friesian		—	United States	—	—	—	—
American Brown Swiss		—	United States	—	—	—	—
American Milking Devon		Taurus	United States	Meat	Dairy	Draught	—
American White Park		Taurus	United States	Meat	Dairy	—	—
Amerifax		Taurus	United States	Meat	Dairy	—	—
Amrit Mahal		Indicus	India	—	—	—	Developed from Hallikar cattle for War purpose
Amsterdam Island cattle		Taurus	Amsterdam Island	—	—	—	Feral
Anatolian Black		Taurus	Turkey	Meat	Dairy	Draught	—
Andalusian Black		Taurus	Spain	Meat	—	—	Endangered
Andalusian Blond		—	Spain	—	—	—	—
Andalusian Grey		—	Spain	—	—	—	—
Angeln cattle		Taurus	Germany	—	Dairy	—	—
Angoni		—	Malawi	—	—	—	—
Angus		Taurus	Scotland	Meat	—	—	—
Ankina		—	United States	—	—	—	—
Ankole-Watusi		Indicus	East Africa	Meat	Dairy	Draught	Show
Aosta cattle		Taurus	Italy	—	—	Draught	—

Alphabetical List - A							
Breed	**Image**	**Subspecies**	**Country/ region of origin**	**Meat**	**Dairy**	**Draught**	**Other**
Apulian Podolian		—	Italy	—	—	—	—
Aracena		—	Spain	—	—	—	—
Arado		—	Ethiopia	—	—	—	—
Argentine Criollo		Taurus	Argentina	Meat	Dairy	—	—
Argentine Friesian		—	Argentine	—	—	—	—
Armorican (cattle)		Taurus	France	Meat	Dairy	—	—
Arouquesa Cattle		Taurus	Portugal	Meat	Dairy	—	—
Arsi		—	Ethiopia	—	—	—	—
Asturian Mountain		Taurus	Spain	Meat	Dairy	—	—
Asturian Valley		Taurus	Spain	Meat	Dairy	—	—
Aubrac cattle		Taurus	France	Meat	Dairy	—	—
Aulie-Ata		Taurus	Kazakhstan	—	Dairy	—	—
Aure et Saint-Girons		Taurus	France	Meat	Dairy	Draught	—
Australian Braford		Taurus/ Indicus hybrid	Australia	Meat	—	—	—

Alphabetical List - A							
Breed	**Image**	**Subspecies**	**Country/ region of origin**	**Meat**	**Dairy**	**Draught**	**Other**
Australian Brangus		Taurus/ Indicus hybrid	Australia	Meat	—	—	—
Australian Charbray		Taurus/ Indicus hybrid	Australia	Meat	—	—	—
Australian Friesian Sahiwal		Taurus/ Indicus hybrid	Australia	—	Dairy	—	—
Australian Milking Zebu		Taurus/ Indicus hybrid	Australia	—	Dairy	—	—
Australian Shorthorn		—	Australia	Meat	—	—	—
Austrian Simmental		—	Austria	—	—	—	—
Austrian Yellow		—	Austria	—	—	—	—
Avetonou		—	Togo	—	—	—	—
Avileña		Taurus	Spain	Meat	—	Draught	—
Avilena-Black Iberian		—	Spain	—	—	—	—
Aweil Dinka		—	South Sudan	—	—	—	—
Ayrshire		Taurus	Scotland	—	Dairy	—	—
Azaouak cattle		—	Mali	—	—	—	—
Azebuado		—	Brazil	—	—	—	—
Azerbaijan Zebu		—	Azerbaijan	—	—	—	—
Azores		—	Portugal	—	—	—	—

B

Alphabetical List - B							
Breed	**Image**	**Subspecies**	**Country/region of origin**	**Meat**	**Dairy**	**Draught**	**Other**
Bachaur cattle		Indicus	India	—	Dairy	Draught	—
Baherie cattle		Indicus	Eritrea	Meat	Dairy	—	—
Bakosi cattle		Taurus	Cameroon	Meat	—	—	Rituals
Balancer cattle		Taurus	United States	Meat	Dairy	—	Gelbvieh/Angus crossbreed
Baoule		Taurus	Ivory Coast	—	—	—	—
Bargur Cattle		Indicus	India	—	Dairy	Draught	Semi wild, native to Bargur forest region, bred and grazed by the local Lingayat community;
Barrosã Cattle		Taurus	Portugal	Meat	—	Draught	—
Barzona		Taurus/ Indicus hybrid	United States	Meat	—	—	—
Bazadais		Taurus	France	Meat	—	—	—
Beef Freisian		Taurus	United States	Meat	Dairy	—	—
Beefalo		Taurus/Bison hybrid	United States	Meat	—	—	—
Beefmaker		Taurus/ Indicus hybrid	United States	Meat	—	—	—
Beefmaster		Taurus/ Indicus hybrid	United States	Meat	—	—	—
Belgian Blue		Taurus	Belgium	Meat	Dairy	—	—
Belgian Red		Taurus	Belgium	Meat	Dairy	Draught	Endangered breed.
Belgian Red Pied		Taurus	Belgium	Meat	Dairy	—	—
Belgian White-and-red		Taurus	Belgium	Meat	Dairy	—	—

Alphabetical List - B							
Breed	**Image**	**Subspecies**	**Country/region of origin**	**Meat**	**Dairy**	**Draught**	**Other**
Belmont Red		Taurus/ Indicus hybrid	Australia	Meat	—	—	—
Belted Galloway		Taurus	Scotland	Meat	—	—	—
Bernese		—	Switzerland	—	—	—	—
Berrenda cattle		Taurus	Spain	Meat	Dairy	Draught	Sport
Betizu		Taurus	Spain/France	Meat	—	Draught	—
Bianca Val Padana		Taurus	Italy	Meat	Dairy	—	—
Blaarkop		Taurus	Netherlands	—	Dairy	—	—
Black Angus		Taurus	Scotland	Meat	Dairy	—	—
Black Baldy		Taurus	Australia	Meat	—	—	—
Black Hereford		Taurus	England	Meat	—	—	—
Blanca Cacereña		Taurus	Spain	Meat	Dairy	Draught	—
Blanco Orejinegro BON		Taurus	Colombia	Meat	Dairy	Draught	—
Blonde d'Aquitaine		Taurus	France	Meat	Dairy	Draught	—
Blue Albion		Taurus	Great Britain	Meat	—	—	—

Alphabetical List - B							
Breed	**Image**	**Subspecies**	**Country/region of origin**	**Meat**	**Dairy**	**Draught**	**Other**
Blue Grey		Taurus	Great Britain	Meat	—	—	—
Bohus Polled		—	Sweden	—	—	—	—
Bonsmara		Taurus/ Indicus hybrid	South Africa	Meat	—	—	—
Boran		Indicus	eastern Africa	Meat	—	—	—
Boškarin		Taurus	Croatia, Slovenia	Meat	Dairy	Draught	—
Braford		Taurus/ Indicus hybrid	United States	Meat	—	—	—
Brahman (cattle)		Indicus	India	Meat	Dairy	Draught	—
Brahmousin		Taurus/ Indicus hybrid	United States	Meat	—	Draught	—
Brangus		Taurus/ Indicus hybrid	United States	Meat	—	—	—
Braunvieh		Taurus	Switzerland	Meat	Dairy	—	—
Brava		—	Spain	—	—	—	—
British White		Taurus	Great Britain	Meat	Dairy	—	—
British Friesian		Taurus	Great Britain	—	Dairy	—	—
Brown Carpathian		Taurus	Ukraine	—	—	—	—

Alphabetical List - B							
Breed	**Image**	**Subspecies**	**Country/region of origin**	**Meat**	**Dairy**	**Draught**	**Other**
Brown Swiss		Taurus	Switzerland	—	Dairy	—	—
Bue Lingo		Taurus	United States	Meat	Dairy	—	—
Burlina		Taurus	Italy	—	Dairy	—	—
Buša cattle		Taurus	former Yugoslavia (Dinaric Alps)	Meat	Dairy	Draught	—
Butana		Indicus	Sudan	—	Dairy	—	—

C

Alphabetical List - C							
Breed	**Image**	**Subspecies**	**Country/region of origin**	**Meat**	**Dairy**	**Draught**	**Other**
Cachena Cattle		Taurus	Portugal/Spain	Meat	Dairy	Draught	—
Caldelana		Taurus	Spain	Meat	Dairy	Draught	—
Camargue		Taurus	France	Meat	—	Draught	Sport
Campbell Island Cattle		Taurus	New Zealand	—	—	—	Feral
Canadian Speckle Park		Taurus	Canada	Meat	—	—	—
Canadienne		Taurus	Canada	Meat	Dairy	—	—
Canaria		Taurus	Spain	Meat	—	Draught	—
Canchim		Taurus/ Indicus hybrid	Brazil	Meat	—	—	—

Alphabetical List - C							
Breed	**Image**	**Subspecies**	**Country/region of origin**	**Meat**	**Dairy**	**Draught**	**Other**
Caracu		Taurus	Brazil	Meat	Dairy	Draught	—
Cárdena andaluza		Taurus	Spain	Meat	—	Draught	—
Carinthian Blondvieh		Taurus	Austria	Meat	Dairy	Draught	—
Carora		Taurus	Venezuelan	Meat	Dairy	Draught	—
Chinese Central Plains Yellow		Taurus/ Indicus hybrid	China	—	—	—	—
Charbray		Taurus/ Indicus hybrid	United States	Meat	—	—	—
Charolais		Taurus	France	Meat	—	Draught	—
Chateaubriand		—	—	—	—	—	—
Chiangus		Taurus	—	Meat	—	—	—
Chianina		Taurus	Italy	Meat	—	Draught	—
Chillingham Cattle		Unknown	England	—	—	—	Feral
Chinese black pied		Taurus	China	—	Dairy	—	—
Cholistani		Indicus	Pakistan	—	Dairy	—	—
Colorsided White Back		Taurus	—	Meat	Dairy	—	—
Commercial		—	—	—	—	—	—

Alphabetical List - C							
Breed	**Image**	**Subspecies**	**Country/region of origin**	**Meat**	**Dairy**	**Draught**	**Other**
Corriente cattle		Taurus	Spain	Meat	Dairy	Draught	Sport
Corsican Cattle		Taurus	Corsica	Meat	—	—	Often free-ranging
Costeño con Cuernos		Taurus/ Indicus hybrid	Colombia	Meat	—	Draught	—
Crioulo Lageano		—	—	—	—	—	—

D

Alphabetical List - D							
Breed	**Image**	**Subspecies**	**Country/ region of origin**	**Meat**	**Dairy**	**Draught**	**Other**
Dajal (cattle)		Indicus	—	Meat	Dairy	Draught	—
Dangi cattle		Indicus	India	—	—	Draught	—
Danish Black-Pied		—	—	—	—	—	—
Danish Jersey		Taurus	Denmark	—	Dairy	—	—
Danish Red		Taurus	Denmark	Meat	Dairy	—	—
Deep Red Cattle		Taurus	—	Meat	Dairy	—	—
Deoni (cattle)		Indicus	India	—	Dairy	Draught	—

Alphabetical List - D							
Breed	**Image**	**Subspecies**	**Country/ region of origin**	**Meat**	**Dairy**	**Draught**	**Other**
Devon		Taurus	England	Meat	Dairy	—	—
Dexter		Taurus	Ireland	Meat	Dairy	—	—
Dhanni (cattle)		Indicus	Pakistan	Meat	Dairy	Draught	—
Doayo cattle		Taurus	Cameroon	—	—	—	—
Doela		—	—	—	—	—	—
Dølafe		Taurus	Norway	Meat	Dairy	—	—
Droughtmaster		Taurus/Indicus hybrid	Australia	Meat	—	—	—
Dulong'		Taurus	—	Meat	—	Draught	—
Dutch Belted		Taurus	Netherlands	Meat	Dairy	—	—
Dutch Friesian		Taurus	Netherlands	Meat	Dairy	—	—
Dwarf Lulu		Taurus/Indicus/ Yak hybrid	—	—	—	—	—

E

Alphabetical List - E							
Breed	**Image**	**Subspecies**	**Country/region of origin**	**Meat**	**Dairy**	**Draught**	**Other**
East Anatolian Red		Taurus	—	—	Dairy	—	—
Eastern Finncattle		Taurus	Finland	Meat	Dairy	—	—
Eastern Red Polled		—	—	—	—	—	—
Enderby Island Cattle		Unknown	New Zealand	—	—	—	Feral
English Longhorn		Taurus	England	Meat	Dairy	—	—
Ennstal Mountain Pied Cattle		Taurus	Austria	Meat	—	Draught	—
Estonian Holstein		—	—	—	—	—	—
Estonian Native		—	—	—	—	—	—
Estonian Red cattle		Taurus	—	Meat	Dairy	—	—
Évolène Cattle		Taurus	Switzerland	—	Dairy	—	—

F

Alphabetical List - F							
Breed	**Image**	**Subspecies**	**Country/region of origin**	**Meat**	**Dairy**	**Draught**	**Other**
Finnish Ayrshire		—	Finland	—	—	—	—
Finnish cattle		Taurus	Finland	Meat	Dairy	—	—
Finnish Holstein-Friesian		—	—	—	—	—	—
Fjäll		Taurus	Sweden	Meat	Dairy	—	Also called Swedish mountain.;
Fleckvieh		Taurus	Austria	Meat	Dairy	—	—
Florida Cracker		Taurus	United States	Meat	—	—	—
French Simmental		Taurus	France	Meat	Dairy	—	—
Fribourg black and white		—	—	—	—	—	—
Friesian Red and White		Taurus	—	Meat	Dairy	—	—
Fulani Sudanese		Indicus	—	Meat	Dairy	Draught	—

G

Alphabetical List - G							
Breed	**Image**	**Subspecies**	**Country/region of origin**	**Meat**	**Dairy**	**Draught**	**Other**
Galician Blond		Taurus	Spain	Meat	Dairy	—	—
Galloway cattle		Taurus	Scotland	Meat	—	Draught	—
Gangatiri cattle		Indicus	India	—	Dairy	Draught	—
Gaolao cattle		Indicus	India	—	Dairy	Draught	—
Garvonesa Cattle		Taurus	Portugal	Meat	—	Draught	—
Gascon cattle		Taurus	France	Meat	—	Draught	—
Gelbvieh		Taurus	Germany	Meat	Dairy	Draught	—
Georgian mountain cattle		Taurus	Georgia	Meat	Dairy	Draught	—
German Angus Cattle		Taurus	Germany	Meat	—	—	—
German Black Pied Cattle		Taurus	Germany	—	Dairy	—	—
German Black Pied Dairy		Taurus	Germany	—	Dairy	—	—
German Red Pied		Taurus	Germany	Meat	Dairy	—	—

Alphabetical List - G							
Breed	**Image**	**Subspecies**	**Country/region of origin**	**Meat**	**Dairy**	**Draught**	**Other**
Gir		Indicus	India	—	Dairy	—	—
Glan cattle		Taurus	Germany	Meat	Dairy	Draught	—
Gloucester		Taurus	England	Meat	Dairy	Draught	—
Gobra		Indicus	—	Meat	—	Draught	—
Greek Shorthorn		Taurus	Greece	Meat	Dairy	—	—
Greek Steppe cattle		Taurus	—	Meat	—	—	—
Greyman Cattle		Taurus/ Indicus hybrid	Australia	Meat	—	—	—
Groningen		Taurus	Netherlands	—	Dairy	—	—
Groningen White-Headed		Taurus	Netherlands	Meat	Dairy	—	—
Gudali		Indicus	—	—	Dairy	Draught	—
Guernsey		Taurus	Channel Islands	—	Dairy	—	—
Guzerat	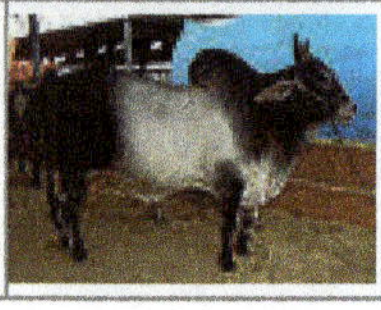	Indicus	India	Meat	—	Draught	—

H

Alphabetical List - H							
Breed	**Image**	**Subspecies**	**Country/ region of origin**	**Meat**	**Dairy**	**Draught**	**Other**
Hallikar		Indicus	India		—	Draught	Origin to Amrit Mahal cattle.
Hanwoo		Taurus	Korea	Meat	—	Draught	Cow fighting
Haryanvi cattle		Indicus	India	—	Dairy	Draught	—
Hartón del Valle		Taurus	Colombia	Meat	Dairy	Draught	—
Harz Red mountain cattle		Taurus	Germany	Meat	Dairy	Draught	—
Hays Converter		Taurus	Canada	Meat	—	—	—
Heck Cattle		Taurus	—	—	—	—	Science
Hereford		Taurus	England	Meat	—	—	—
Herens		Taurus	Switzerland	Meat	—	—	Cow fighting
Hhybridmaster		Taurus/ Indicus hybrid	—	Meat	Dairy	—	—
Highland Cattle		Taurus	Scotland	Meat	—	—	—

Alphabetical List - H							
Breed	**Image**	**Subspecies**	**Country/ region of origin**	**Meat**	**Dairy**	**Draught**	**Other**
Hinterwald Cattle		Taurus	Germany	Meat	Dairy	—	—
Holando-Argentino		Taurus	Argentina	Meat	Dairy	—	—
Red Holstein		Taurus	—	—	Dairy	—	—
Holstein		Taurus	—	—	Dairy	—	—
Horro		Indicus	—	Meat	—	Draught	—
Hungarian Grey		Taurus	Hungary	Meat	—	Draught	—

I

Alphabetical List - I							
Breed	**Image**	**Subspecies**	**Country/region of origin**	**Meat**	**Dairy**	**Draught**	**Other**
Iberian cattle		—	—	—	—	—	—
Icelandic		Taurus	Iceland	—	Dairy	—	—
Illawarra cattle		Taurus	Australia	Meat	Dairy	—	—
Improved Red and White		Taurus	—	Meat	—	—	—

Alphabetical List - I							
Breed	**Image**	**Subspecies**	**Country/region of origin**	**Meat**	**Dairy**	**Draught**	**Other**
Indo-Brazilian		Indicus	—	Meat	—	—	—
Irish Moiled		Taurus	Ireland	Meat	Dairy	—	—
Israeli Holstein		Taurus	—	—	Dairy	—	—
Israeli Red		Taurus/ Indicus hybrid	—	Meat	Dairy	—	—
Istoben cattle		Taurus	—	Meat	Dairy	—	—

J

Alphabetical List - J							
Breed	**Image**	**Subspecies**	**Country/ region of origin**	**Meat**	**Dairy**	**Draught**	**Other**
Jamaica Black		Taurus/Indicus hybrid	—	Meat	Dairy	—	—
Jamaica Hope		Taurus/Indicus hybrid	Jamaica	—	Dairy	—	—
Jamaica Red		Taurus/Indicus hybrid	—	Meat	—	—	—
Jarmelista Cattle		Taurus	Portugal	Meat	—	Draught	—
Javari Cattle		Indicus	India	—	—	Draught	—
Jersey		Taurus	Channel Islands	—	Dairy	—	—
Jutland cattle		Taurus	Denmark	Meat	Dairy	—	—

K

Alphabetical List - K							
Breed	**Image**	**Subspecies**	**Country/ region of origin**	**Meat**	**Dairy**	**Draught**	**Other**
Kabin Buri cattle		Taurus/Indicus hybrid	Thailand	Meat	Dairy	—	—
Kalmyk cattle		Taurus	—	Meat	Dairy	—	—
Kangayam		Indicus	India	—	—	Draught	—
Kankrej	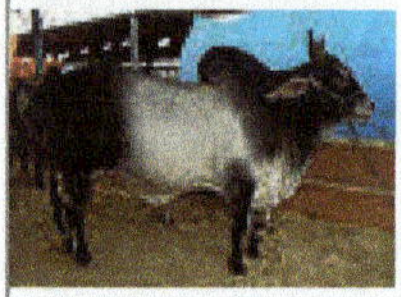	Indicus	India	—	—	Draught	—
Kamphaeng Saen cattle		Taurus/Indicus hybrid	Thailand	Meat	—	—	—
Karan Swiss		Taurus/Indicus hybrid	—	Meat	Dairy	—	—
Kasaragod Dwarf cattle		Indicus	India	—	Dairy	Draught	—
Kathiawadi		Indicus	India	—	—	Draught	—
Kazakh Whiteheaded		Taurus	—	Meat	Dairy	—	—
Kenana		Indicus	Sudan	—	Dairy	—	—
Kenkatha cattle		Indicus	India	—	—	Draught	—
Kerry cattle		Taurus	Ireland	Meat	Dairy	—	—
Kherigarh cattle		Indicus	India	—	—	Draught	—

Alphabetical List - K							
Breed	**Image**	**Subspecies**	**Country/ region of origin**	**Meat**	**Dairy**	**Draught**	**Other**
Khillari cattle		Indicus	India	—	—	Draught	—
Kholomogory		Taurus	Russia	Meat	Dairy	—	—
Korat Wagyu		Taurus/Indicus hybrid	Thailand	Meat	—	—	—
Kostroma Cattle		Taurus	—	Meat	Dairy	—	—
Krishna Valley cattle		Indicus	India	—	Dairy	Draught	—
Kuri		Taurus	—	Meat	Dairy	Draught	—
Kurgan		Taurus	—	Meat	Dairy	—	—

L

Alphabetical List - L							
Breed	**Image**	**Subspecies**	**Country/ region of origin**	**Meat**	**Dairy**	**Draught**	**Other**
Lampurger		—	—	—	—	—	—
Latvian Blue		—	—	—	—	—	—
Latvian Brown		Taurus	Latvia	Meat	Dairy	—	—
Lebedyn (cattle)		Taurus	Ukraine	—	—	—	—
Levantina		Taurus	—	Meat	—	Draught	—
Limia Cattle		Taurus	Spain	Meat	Dairy	Draught	—

Alphabetical List - L							
Breed	**Image**	**Subspecies**	**Country/ region of origin**	**Meat**	**Dairy**	**Draught**	**Other**
Limousin		Taurus	France	Meat	—	Draught	—
Limpurger		Taurus	—	Meat	Dairy	—	—
Lincoln Red		Taurus	England	Meat	—	—	—
Lineback		Taurus	—	—	Dairy	—	—
Lithuanian Black-and-White		—	—	—	—	—	—
Lithuanian Light Grey		—	—	—	—	—	—
Lithuanian Red		—	Lithuania	Meat	—	—	—
Lithuanian White-backed		—	—	—	—	—	—
Lohani cattle		Indicus	—	Meat	—	Draught	—
Lourdais		Taurus	France	Meat	Dairy	Draught	—
Australian Lowline		Taurus	Australia	Meat	—	—	—

Alphabetical List - L							
Breed	**Image**	**Subspecies**	**Country/ region of origin**	**Meat**	**Dairy**	**Draught**	**Other**
Luing		Taurus	Scotland	Meat	—	—	—

M

Alphabetical List - M							
Breed	**Image**	**Subspecies**	**Country/ region of origin**	**Meat**	**Dairy**	**Draught**	**Other**
Madagascar Zebu		Indicus	—	Meat	Dairy	Draught	—
Madura cattle		Banteng/ Indicus hybrid	Indonesia	Meat	Dairy	Draught	Racing
Maine Anjou		Taurus	France	Meat	Dairy	Draught	—
Malnad Gidda		Indicus	India	—	Dairy	Draught	Semi dwarf breed
Malvi		Indicus	India	—	—	Draught	—
Mandalong Special		Taurus/ Indicus hybrid	—	Meat	—	—	—
Mantequera Leonesa		Taurus	—	Meat	Dairy	Draught	—
Maramureş Brown		—	Romania	—	—	—	—

Alphabetical List - M							
Breed	**Image**	**Subspecies**	**Country/ region of origin**	**Meat**	**Dairy**	**Draught**	**Other**
Marchigiana		Taurus	Italy	Meat	—	Draught	—
Maremmana		Taurus	Italy	Meat	Dairy	Draught	—
Marinhoa Cattle		Taurus	Portugal	Meat	Dairy	Draught	—
Maronesa		Taurus	Portugal	Meat	Dairy	Draught	—
Masai cattle		Indicus	—	Meat	—	Draught	—
Mashona		Taurus	—	Meat	—	Draught	—
Menorquina		Taurus	—	Meat	—	Draught	—
Mertolenga Cattle		Taurus	Portugal	Meat	—	Draught	—
Meuse-Rhine-Issel		Taurus	Netherlands	Meat	Dairy	—	—
Mewati cattle		Indicus	India	—	Dairy	Draught	—
Milking Shorthorn		Taurus	Great Britain	Meat	Dairy	—	—
Minhota Cattle		Taurus	Portugal	Meat	Dairy	Draught	—
Miniature		—	—	—	—	—	—

Alphabetical List - M							
Breed	**Image**	**Subspecies**	**Country/ region of origin**	**Meat**	**Dairy**	**Draught**	**Other**
Mirandesa cattle		Taurus	Portugal	Meat	—	Draught	—
Mocăniță (cattle)		—	Romania	—	—	—	—
Mollie (cattle)		Taurus	Spain	Meat	—	Draught	—
Monchina (cattle)		Taurus	Spain	Meat	—	Draught	—
Mongolian cattle		Taurus	China	Meat	—	Draught	—
Montbeliard Cattle		Taurus	France	Meat	Dairy	—	—
Morucha		Taurus	—	Meat	—	Draught	Fighting
Murboden Cattle		Taurus	Austria	Meat	Dairy	Draught	—
Murnau-Werdenfels Cattle		Taurus	Germany	—	Dairy	—	—
Murray Grey		Taurus	Australia	Meat	—	—	—

N

Alphabetical List - N							
Breed	**Image**	**Subspecies**	**Country/ region of origin**	**Meat**	**Dairy**	**Draught**	**Other**
Nagori		Indicus	India	—	—	Draught	—
N'Dama		Taurus	Guinea	Meat	Dairy	—	—
Negra Andaluza		Taurus	Spain	Meat	—	Draught	—
Nelore cattle		Indicus	India	Meat	—	Draught	—
Nguni		Taurus/ Indicus hybrid	—	Meat	Dairy	—	—
Nimari cattle		Indicus	India	—	—	Draught	—
Normande Cattle		Taurus	France	Meat	Dairy	—	—

Alphabetical List - N							
Breed	**Image**	**Subspecies**	**Country/ region of origin**	**Meat**	**Dairy**	**Draught**	**Other**
Northern Finncattle		Taurus	Finland	Meat	Dairy	—	—
Northern Shorthorn		Taurus	Great Britain	—	Dairy	—	—
Chinese Northern Yellow		—	China	—	—	—	—
Norwegian Red		Taurus	Norway	Meat	Dairy	—	—

LIST OF CHICKEN BREEDS

There are hundreds of chicken breeds in existence. Domesticated for thousands of years, distinguishable breeds of chicken have been present since the combined factors of geographical isolation and selection for desired characteristics created regional types with distinct physical and behavioral traits passed on to their offspring.

The physical traits used to distinguish chicken breeds are size, plumage color, comb type, skin color, number of toes, amount of feathering, egg color, and place of origin. They are also roughly divided by primary use, whether for eggs, meat, or ornamental purposes, and with some considered to be dual-purpose.

In the 21st century, chickens are frequently bred according to predetermined breed standards set down by governing organizations. The first of such standards was the British Poultry Standard, which is still in publication today. Other standards include the Standard of Perfection, the Australian Poultry Standard, and the standard of the American Bantam Association, which deals exclusively with bantam fowl. Only some of the known breeds are included in these publications, and only those breeds are eligible to be shown competitively. There are additionally a few hybrid strains which are common in the poultry world, especially in large poultry farms. These types are first generation crosses of true breeds. Hybrids do not reliably pass on their features to their offspring, but are highly valued for their producing abilities.

Thirty-nine varieties of chicken (and one Guinea Fowl).

LIST BY PLACE OF ORIGIN

Afghanistan

• Ingriido • Khasaki • Kulangi	• Pusty • Sabzwari

Albania

• Commune hen • Black Tropoja Lekbibaj • Tirana	• Tropoja Pac • Yerevanian

Australia

The Australorp, an Australian breed

• Australorp	• Australian Game
• Australian Langshan	• Australian Pit Game

Austria

• Altsteirer	• Sulmtaler

Belgium

The d'Everberg, a Belgian ornamental bantam

- Ardennaise or Ardenner (large and bantam)
- Ardenner Bolstaart
- Belgian Bearded d'Anvers, Antwerp Belgian, Barbu d'Anvers or Antwerpse baardkriel (bantam)
- Barbu de Boitsfort or Bosvoordse baardkriel (bantam)
- Barbu de Grubbe or Grubbe baardkriel (bantam)
- Belgian d'Everberg, Barbu d'Everberg or Everbergse baardkriel (bantam)
- Barbu de Watermael or Watermaalse baardkriel (bantam)
- Bearded d'Uccle, Barbu d'Uccle or Ukkelse baardkriel (bantam)
- Bassette Liègeoise (bantam)
- Brabanter, Brabançonne or Brabants hoen (large and bantam)
- Braekel or Brakel (large and bantam)
- Campine
- Combattant de Bruges or Brugse vechter (large and bantam)
- Combattant de Liège or Luikse vechter (large and bantam)
- Combattant de Tirlemont or Tiense vechter (large and bantam)
- Coucou d'Izegem or Izegemse koekoek
- Coucou des Flandres or Vlaanderse koekoek
- Famennoise or Famennehoen (large and bantam)
- Fauve de Hesbaye or Gele van Haspengouw
- Fauve de Mehaigne or Gele van Mehaigne (bantam)
- Herve or Hervehoen (large and bantam)
- Huttegem
- Malines or Mechels hoen (large and bantam)
- Mechelse kalkoenkop
- Naine belge or Belgische kriel (bantam)
- Naine de Waes or Waasse kriel (bantam)
- Naine du Tournaisis or Doornikse kriel (bantam)
- Pondeuse de Zingem or Zingems leghoen
- Poule d'Aarschot or Aarschots hoen
- Poule de la vallée de la Zwalm or Hoen van de Zwalmvallei
- Poule de Zottegem or Zottegems hoen
- Zingems vleeshoen

Bulgaria

Black Shumen cock

• Black Shumen	• Starozagorska Red

Canada

• Chantecler	• Red Shaver

Chile

- Araucana

China

A Silkie hen

• Cochin • Croad Langshan • Nankin	• Pekin • Silkie

PeKin Oil Chicken

Pekin Oil Chicken, namely, China Court Yellow Chicken,is a breed own about more than 300 years history in Bei Jing, the Capital of China.

Croatia

- Hrvatica (Croatian Hen, Kokoš Hrvatica)

Cuba

Cubalaya cock

- Cubalaya

Cyprus

- Cypriot chicken

Czech Republic

• Czech gold brindle chicken (Česká slepice zlatě kropenatá, Česká zlatá kropenka, Češka)	• Šumavanka

Egypt

• Alexandria	• Golden Montazah
• Arbrikers	• Hawara
• Bahig	• Inshas
• Bandara	• Iven
• Boffen	• Kherbis
• Dandarawi	• Mandarah
• Dokki 4	• Matrouh
• Baladi Beheri	• Salam
• Fayoumi	• Silver Montazah
• Gimmizah	• Sinai

Finland

• Alho	• Jussila
• Finnish Landrace chicken (Finnish chicken)	• Kiuruvesi
• Häme	• Lindellin
• Hornio	• Luumäki
• Iitti	• Piikkiö
• Ilmajoki	• Savitaipale
• Jalasjärvi	• Tyrnävä

France

Crèvecœur

La Flèche

Faverolles

- Aquitaine
- Ardeale
- Ardennaise, or Sans-queue des Ardennes
- Barbezieux
- Blanzac
- Bourbonnaise
- Bourbourg
- Bresse Gauloise
- Caumont
- Caussade
- Charollaise
- Cocherelle
- Combattant du Nord
- Contres
- Cotentine
- Coucou des Flandres
- Coucou de France
- Coucou de Rennes
- Coucou Picarde
- Coucou Soie
- Cou-nu du Forez
- Courtes-Pattes
- Crèvecœur
- Estaires
- Faverolles clair (light)
- Faverolles foncée (dark or German Faverolles)
- Favoris
- Gasconne
- Gâtinaise
- Gauloise dorée
- Géline de Touraine
- Gournay
- Hergnies
- Houdan
- Ivanaise
- Janzé
- Javanaise
- La Flèche
- Landaise
- Le Mans
- Le Merlerault
- Meusienne
- Limousine (coq de pêche)
- Lyonnaise
- Mantes
- Marans
- Noire du Berry
- Noire de Challans
- Pavilly
- Poule d'Alsace
- Poule de Caux
- Poule de Marquise
- Poule de Saint-Omer
- Poule des Courrières
- Poule des haies, see Ardennaise
- Provençale

Germany

The oldest German breed, the Bergische Kräher (Historical illustration by Jean Bungartz, 1885)

The longtailed Phoenix, a German breed derived from the Japanese Onagadori

The auto-sexing Bielefelder, created in the 70s

Large Breeds

- Annaberger Haubenstrupphühner
- Augsburger
- *Bergische Kräher*,
- Bergische Schlotterkämme
- Bielefelder Kennhuhn
- *Deutsches Lachshuhn*, see Faverolles
- *Deutsche Langschan*, see German Langshan
- Deutsches Reichshuhn
- Deutsche Sperber
- Dresdner
- *Hamburger*, see Hamburg
- Kraienköppe (*Twents hoen*)
- Krüper
- *Lakenfelder*, see Lakenvelder
- *Nackthalshühner*, see Naked-neck
- Niederrheiner
- *Ostfriesische Möwen*, see East Frisian Gull
- Pfälzer Kämpfer
- Phoenix
- Ramelsloher
- Rheinländer
- Sachsenhuhn
- *Strupphuhn*, see Frizzle
- *Sultanhuhn*, see Sultan
- Sundheimer
- Thüringer Barthuhn
- Vogtländer
- *Vorwerkhuhn*, see Vorwerk
- *Westfälische Totleger*, see Westphalian chicken

Bantams

- Bergische Zwerg-Kräher
- Bergische Zwerg-Schlotterkämme
- Bielefelder Zwerg-Kennhühner
- Brügger Zwerg-Kämpfer
- Deutsche Zwerghühner
- *Deutsche Zwerg-Lachshühner*, see Faverolle Bantam
- *Deutsche Zwerg-Langschan*, see German Langshan Bantam
- Deutsche Zwerg-Reichshühner
- Deutsche Zwerg-Sperber
- Frankfurter Zwerghühner
- Ostfriesische Zwerg-Möwen

- Ruhlaer Zwerg-Kaulhühner
- Siamesisches Zwerg-Seidenhühner
- Thüringer Zwerg-Barthühner
- Zwerg-Altsteirer
- Zwerg-Amrocks
- *Zwerg-Andalusier*, see Andalusian
- *Zwerg-Araucana*, see Rumpless Araucana Bantam
- *Zwerg-Asil*, see Asil Bantam
- *Zwerg-Augsburger*, see Augsburger Bantam
- *Zwerg-Australorps*, see Australorp Bantam
- *Zwerg-Barnevelder*, see Barnevelder Bantam
- *Zwerg-Brahma*, see Brahmas
- *Zwerg-Brakel*, see Brakel Bantam
- *Zwerg-Créve Coeur*, see Crevecoeur Bantam
- *Zwerg-Dominikaner*, see Dominique Bantam
- Zwerg-Dresdner
- *Zwerg-Hamburger*, see Hamburg Bantam
- *Zwerg-Italiener*, see Leghorn Bantam
- Zwerg-Kastilianer
- Zwerg-Krüper
- *Zwerg-La Fléche*, see La Fleche Bantam
- *Zwerg-Lakenfelder*, see Lakenfelder Bantam
- *Zwerg-Minorka*, see Minorca Bantam
- *Zwerg-Nackthalshühner*, see Naked-neck bantam
- *Zwerg-New Hampshire*, see New Hampshire Bantam
- Zwerg-Niederrheiner
- *Zwerg-Orloff*, see Orloff Bantam
- *Zwerg-Orpinglon* see Orpington Bantam
- *Zwerg-Phönix*, see Phoenix
- *Zwerg-Plymouth Rocks*, see Plymouth Rock Bantam
- Zwerg-Rheinländer
- Zwerg-Rhodeländer
- Zwerg-Sachsenhühner
- *Zwerg-Sulmtaler*, see Sulmtaler Bantam
- Zwerg-Sundheimer
- *Zwerg-Sussex*, see Sussex Bantam
- *Zwerg-Vorwerkhühner* see Vorwerk Bantam
- *Zwerg-Welsumer* see Welsumer Bantam
- *Zwerg-Yokohama* see Yokohama Bantam

Greece

• Alonissos island chicken • Chios Fighting chicken • Follidotes chicken • Kalamata chicken	• Pomak Fighting chicken • Lesvos Fillianes chicken • Trikala chicken

Iceland

Icelandic chicken

India

The Aseel, a fighting breed

• Ankaleshwar	• Kadaknath
• Aseel	• Kalasthi
• Busra	• Kalinga Brown
• Cari Gold	• Kashmir Faverolla
• Danki	• Krishna-J
• Daothigir	• Miri
• Debendra	• Mrityunjay
• Dhanraja	• Nicobari
• Ghagus	• Punjab Brown
• Giriraja	• Tellicherry
• Gramalakshmi	• Vanaraja
• Gramapriya	• Yamuna
	• kushi

Indonesia

Sumatra cock

• Ayam Cemani • Bekisar (interspecific hybrid) • Ketawa • Sumatra chicken	• Kedu (nationally standardized) • Sentul (nationally standardized) • Kokok Balenggek (nationally standardized) • Pelung (long-crower from West Java, locally standardized)

Iran

The Orloff, an Iranian breed

• Manx Rumpy (or Rumpless Game)	• Orloff

Italy

Leghorn hen

• Ancona	• Padovana
• Bianca di Saluzzo	• Padovana Riccia
• Bionda Piemontese	• Pepoi
• Ciuffine Ghigi	• Polverara
• Collo Nudo Italiano	• Robusta Lionata
• Ermellinata di Rovigo	• Robusta Maculata
• Livorno, see Leghorn	• Romagnola
• Mericanel della Brianza	• Siciliana
• Millefiori di Lonigo	• Sicilian Buttercup
• Millefiori Piemontese	• Valdarno
• Modenese	• Valdarnese
• Mugellese	

Japan

• Japanese Bantam (or Chabo) • Shamo (or Ko-Shamo)	• Tomaru • Onagadori

Korea

- Jangmigye

Kosovo

- Kosova Long-crowing

Malaysia

• Malay	• Serama

Marianas

- Saipan Jungle Fowl

Myanmar

- Burmese

Netherlands

The Barnevelder

• Assendelfts Hoen (large and bantam) • Barnevelder (large and bantam) • Brabanter (large and bantam) • Chaamse hoenders • Drents Hoen (large and bantam) • Eikenburgerkriel (bantam) • Fries Hoen (large and bantam) • Friesian • Groninger Meeuw (large and bantam)	• Hollandse Kriel, see Dutch Bantam (bantam) • Hollandse Kuifhoen (large and bantam) • Hollands Hoen (large and bantam), see Hamburg chicken (large and bantam) • Kraaikop (large and bantam) • Lakenvelder (large and bantam) • Nederlandse Baardkuifhoen (large and bantam) • Nederlandse Sabelpootkriel, see Dutch Booted Bantam (bantam) • Nederlandse Uilebaard (large and bantam) • Noord Hollands Hoen or Noord-Hollandse Blauwe, see North Holland Blue (large and bantam) • Schijndelaar • Twents Hoen (large and bantam) • Welsummer (large and bantam)

Norway

- Norwegian Jærhøne

Pakistan

• Asil (or Aseel) • Sindhi	• Lyallpur Silver black

Philippines

- Banaba
- Bolinao
- Camarines
- Darag
- Paraoakan or Parawakan

Poland

• Polbar	• Zielononóżka Kuropatwiana (Green-legged Partridge)

Portugal

• Amarela	• Pedrês Portuguesa
• Preta Lusitânica	• Raça Branca

Romania

Naked-neck

- Transylvanian Naked-neck

Russia

• Russian Orloff	• Yurlov Crower

Serbia

• Banat Naked-neck (Banatski gološijan) • Sombor Crested chicken (Somborska kaporka)	• Svrljig Black (Svrljiška kokoš)

Slovakia

- Oravka

South Africa

• Ovambo • Potchefstroom Koekoek	• Venda • Boschveld

Spain

A Minorca rooster

A Cantabrian Pedresa rooster without tail feathers

• Andalusian, Andaluza Azul	• Murciana
• Cara Blanca	• Galiña de Mos
• Castilian, Castellana Negra	• Pardo de León
• Catalana, Catalana Del Prat or Prat Leonada	• Penedesenca
• Empordanesa	• Pedresa, Cuca, Franciscana or Cantabrian
• Euskal oiloa	• Pita Pinta Asturiana
• Extremaduran, Extremeña Azul	• Pitiüsa
• Indio de León	• Spanish game, Combatiente español
• Gallina Mallorquina	• Utrerana
• Minorca, Gallina Menorquina	• Valenciana de Chulilla
	• White-Faced Black Spanish

Sweden

• Åsbohöna • Bohuslän-Dals svarthöna • Gotlandshöna • Hedemorahöna • Ölandshöns	• Öländsk dvärghöna • Orusthöna • Skånsk blommehöna • Svensk dvärghöna • Silverudd Blue/ Isbar

Switzerland

The Appenzeller, a rare Swiss crested breed

• Appenzeller Barthühner, see Appenzeller • Appenzeller Spitzhauben, see Appenzeller • Schweizerhuhn	• Zwerg-Appenzeller Barthuhn, see Appenzeller • Zwerg-Schweizerhuhn, see Schweizer chicken

Thailand

- Gai Chon
- Gai Puen Muang
- Pradu Hang Dum or Pradu Hang Dam Chiangmai
- Thai Game

Turkey

The Sultan, a Turkish breed

- Abaza
- Denizli
- Gebze
- Gerze
- Hacikadin
- Sultan

Ukraine

- Poltava

United Kingdom

A Buff Orpington cock

• Burford Brown • Cornish (sometimes called the "Indian Game" although not from India) • Derbyshire Redcap • Dorking • Ixworth • Marsh Daisy • Modern Game • Modern Langshan • Muffed Old English Game • Norfolk Grey	• Old English Game • Old English Pheasant Fowl • Orpington • Rosecomb • Scots Dumpy • Scots Grey • Sebright • Sussex

United States of America

• Ameraucana	• Jersey Giant
• American Game	• Lamona
• Buckeye	• New Hampshire
• Blue Hen of Delaware	• Plymouth Rock (or Barred Rock, Rock)
• California Gray	• Pyncheon
• Delaware	• Rhode Island Red
• Dominique	• Rhode Island White
• Holland	• Winnebago
• Iowa Blue	• Wyandotte
• Java	• Leghorn chicken

Vietnam

• Ga Ac	• Ca To
• Ga Choi	• Ga Tre Tan Chau
• Ga Dong Tao	• H'mong
• Gà Hồ	• Oke
• Ga Mia	• Phu lu Te
• Ga Mong	• Te
• Ga Noi	• Tien Yen
• Ga Ri	• Vanphu
• Ga Tau Vang	

By Primary Use

All chickens lay eggs, have edible meat, and possess a unique appearance. However, distinct breeds are the result of selective breeding to emphasize certain traits. Any breed may technically be used for general agricultural purposes, and all breeds are shown to some degree. But each chicken breed is known for a primary use.

Eggs

An Araucana

Many breeds were selected and are used primarily for producing eggs, these are mostly light-weight birds whose hens do not go broody often.

• Ancona • Andalusian • Asturian Painted Hen • Barnevelder • Campine • Catalana • Easter Egger • Egyptian Fayoumi • Norwegian Jærhøne	• Kraienköppe (Twentse) • Lakenvelder • Leghorn • Marans • Minorca • Orloff • Penedesenca • Sicilian Buttercup • White-Faced Black Spanish • Welsummer

Meat

Most farms and homesteads use dual-purpose breeds for meat production. Some breeds are raised mainly for meat:

• Bresse • Cornish (a.k.a. Indian Game)	• Ixworth • Jersey Giant

Dual-Purpose

Barred Plymouth Rock, a dual-purpose breed

The generalist breeds used in barnyards worldwide are adaptable utility birds good at producing both meat and eggs. Though some may be slightly better for one of these purposes, they are usually called *dual-purpose* breeds.

• Australorp	• Marsh Daisy
• Brahma	• Naked-neck
• Braekel (Brakel)	• New Hampshire
• Buckeye	• Norfolk Grey
• California Gray	• Orpington
• Chantecler	• Plymouth Rock
• Cubalaya	• Poltava
• Derbyshire Redcap	• Red Shaver
• Dominique	• Rhode Island Red
• Dorking	• Rhode Island White
• Faverolles	• Scots Dumpy
• Holland	• Scots Grey
• Iowa Blue	• Sussex
• Java	• Winnebago
• Jersey Giant	• Wyandotte

Exhibition

Since the 19th century, poultry fancy, the breeding and competitive exhibition of poultry as a hobby, has grown to be a huge influence on chicken breeds. Many breeds have always been kept for ornamental purposes, and others have been shifted from their original use to become first and foremost exhibition fowl, even if they may retain some inherent utility. Since the sport of cockfighting has been outlawed in the developed world, most breeds first developed for this purpose, called game fowl, are now seen principally in the show ring rather than the cock pit as fighting cocks.

Modern Game Fowl

• American Game G	• Jangmigye
• Asil G	• La Flèche U
• Appenzeller U	• Malay G
• Barbu de Watermael	• Modern Game
• Belgian Bearded d'Anvers	• Nankin
• Bearded d'Uccle	• Old English Game G
• Belgian d'Everberg	• Pekin
• Blue Hen of Delaware G	• Phoenix
• Booted Bantam	• Polish U
• Cochin	• Rosecomb
• Crèvecœur U	• Sebright
• Croad Langshan U	• Serama
• Dutch Bantam	• Shamo G
• Frizzle	• Silkie
• Ga NoiG	• Sultan
• Hamburg U	• Sumatra
• Houdan U	• Vorwerk U
• Japanese Bantam	• Yokohama

Bantams

Golden Sebright cockerel

Most large chicken breeds have a bantam counterpart, sometimes referred to as a *miniature*. Miniatures are usually one-fifth to one-quarter the size of the standard breed, but they are expected to exhibit all of the standard breed's characteristics. A *true bantam* has no large counterpart, and is naturally small. The true bantams include:

• Belgian Bearded d'Anvers • Belgian Bearded d'Uccle • Belgian d'Everberg • Booted Bantam • Dutch Bantam • Japanese Bantam	• Nankin • Pekin • Rosecomb • Sebright • Serama • Silkie

Cross-Breeds

Many common strains of cross-bred chickens exist, but none breed true or are recognized by poultry breed standards. Thus, though they are extremely common in flocks focusing on high productivity, cross-breeds do not technically meet the definition of a breed. Most cross-breed strains are sex linked, allowing for easy chick sexing.

• Black Sex Link (also called Black Stars or * Black Rock in the UK) • Red Sex Link (also called Red Stars) • ISA Brown • Kuroiler • Lohmann Brown	• Daisy Belle • Cream Legbar • Cornish-Rock • Easter Egger • Broiler

DOG BREEDING

Dog breeding is the practice of mating selected dogs with the intent to maintain or produce specific qualities and characteristics. When dogs reproduce without such human intervention, their offsprings' characteristics are determined by natural selection, while "dog breeding" refers specifically to the artificial selection of dogs, in which dogs are intentionally bred by their owners. A person who intentionally mates dogs to produce puppies is referred to as a dog breeder. Breeding relies on the science of genetics, so the breeder with a knowledge of canine genetics, health, and the intended use for the dogs attempts to breed suitable dogs.

A litter of puppies and their mother

History

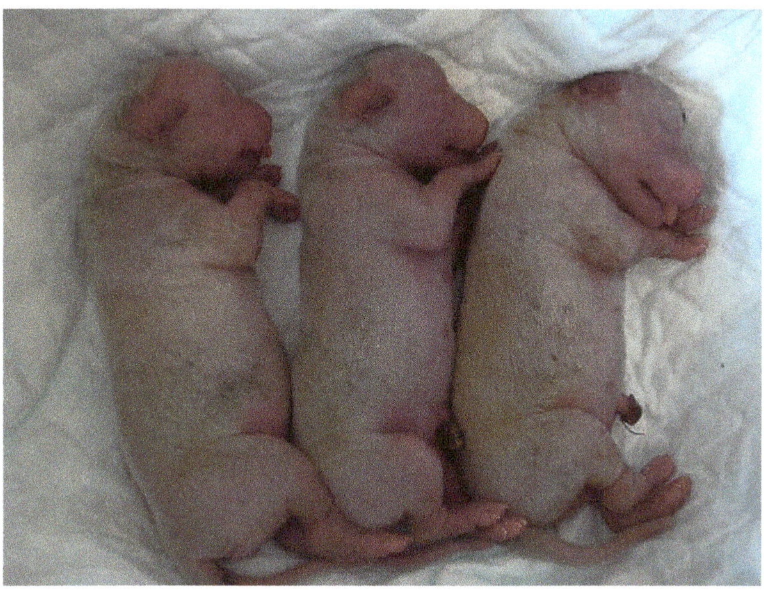

Few hours old

Three days old

Few weeks old

Seven weeks old

Twelve weeks old

Humans have maintained populations of useful animals around their places of habitat since pre-historic times. Over these millennia, domesticated dogs have developed into distinct types, or *groups*, such as livestock guardian dogs, hunting dogs, and sighthounds. Artificial selection in dog breeding has influenced the dog's behavior, shape, and size.

Some researchers believe that the evolution of dogs from wolves is an example of neoteny or paedomorphism selection, which results in a retention of juvenile physical characteristics. Compared with wolves, many adult dog breeds retain such juvenile characteristics as soft fuzzy fur, round torsos, large heads and eyes, ears that hang down rather than stand erect, etc.; characteristics that are shared by most juvenile mammals, and therefore generally elicit some degree of protective and nurturing behavior cross-species from most adult mammals, including humans, who term such characteristics "cute" or "appealing". It has been seen that these traits can even prompt an adult female wolf to act more defensively of dog puppies than of wolf puppies. The example of canine neoteny goes even further, in that the various dog breeds are differently neotenized according to the type of behavior that was selected. Other researchers believe that because this comparison is based on the gray wolf, which is not the ancestor of the dog, that this comparison is invalid. Further research indicates that the concept of neoteny as a means of distinguishing dogs from wolves is baseless.

To maintain these distinctions, humans have intentionally mated dogs with certain characteristics to encourage those characteristics in the offspring. Through this process, hundreds of dog breeds have been developed. Initially, the ownership of working and, later, purebred dogs, was a privilege of the wealthy. Nowadays, many people can afford to buy a dog. Some breeders chose to breed purebred dogs, while some prefer the birth of a litter of puppies to a dog registry, such as kennel club to record it in stud books such as those kept by the AKC (American Kennel Club). Such registries maintain records of dogs' lineage and are usually affiliated with kennel clubs. Maintaining correct data is important for purebred dog breeding. Access to records allows a breeder to analyze the pedigrees and anticipate traits and behaviors. Requirements for the breeding of registered purebreds vary between breeds, countries, kennel clubs and registries. It has been concluded that "findings imply that when selective breeding was done by humans, it squashed the snouts of certain dog breeds, it also morphed their brains" (Scientific American, 2010). Breeders have to abide by the rules of the specific organization to participate in its breed maintenance and development programs. The rules may apply to the health of the dogs, such as joint x-rays, hip certifications, and eye examinations; to working qualities, such as passing a special test or achieving at a trial; to general conformation, such as evaluation of a dog by a breed expert. However, many registries, particularly those in North America, are not policing agencies that exclude dogs of poor quality or health. Their main function is simply to register puppies born of parents who are themselves registered.

Criticism

Some dogs have certain inheritable characteristics that can develop into a disability or disease. Canine hip dysplasia is one such condition. Some eye abnormalities, some heart conditions, and some cases of deafness have been proven to be inherited. There have been extensive studies of these conditions, commonly sponsored by breed clubs and dog registries, while specialised breed clubs provide information of common genetic defects for their breeds. As well, special organizations, such as Orthopedic Foundation for Animals, collect data and provide it to breeders, as well as to the general public. Conditions such as hip dysplasia can impact some breeds more than others.

Some registries, such as American Kennel Club, may include a record of the absence of certain genetic defects, known as a certification, in an individual dog's record. For example, the German Shepherd Dog national breed club in Germany is a registry that recognizes that hip dysplasia is a genetic defect for dogs of this breed. Accordingly, it requires all dogs to pass evaluation for absence of hip dysplasia to register their progeny, and records the results in individual dog's pedigrees.

There are BBC documentaries titled "Pedigree Dogs Exposed" and "Pedigree Dogs Exposed – Three Years On" that claim health problems in dogs resulting from inbreeding. Problems such as breathing in the Pug breed and Pekingese breed, spinal problems in the Dachshund breed, and Syringomyelia in the Cavalier King Charles Spaniel breed.

Some scientific researchers argue that advances in artificial reproduction technology for the purposes of dog breeding can be helpful but also have "detrimental impacts" when overused instead of natural selection principles. These scientists call for a deeper understanding of natural selection, leading to a more naturalistic approach in dog breeding.

Inbreeding Depression

Inbreeding depression is the reduced survival and fertility of offspring of related individuals. A study of 42,855 dachshund litters found that as the inbreeding coefficient increased, litter size decreased and the percentage of stillborn puppies increased, thus indicating inbreeding depression.

LIST OF DONKEY BREEDS

This list of breeds of domestic donkey is based on country reports to the international DAD-IS database.

Breeds

Breed name English/ (*native language in DAD-IS*) /Other names	**Country**	**Notes**	**Image**
Abkhazskaya	Russian Federation		
Abyssinian Donkey	Ethiopia		
Algerian	Algeria	Used as a draught or meat animal. Chestnut or grey in color.	

American Mammoth Jack	USA	Also called "Mammoth Jackstock" or "Mammoth donkey"; developed from a blend of Poitou, Andalusian, Majorcan, and Catalan breeds	
Amiatina (*Asino dell'Amiata*)	Italy	Monte Amiata, province of Grosseto, Tuscany	
Anatolian	Turkey		
Andalusian donkey (*Asno Andaluz*)	Spain		
Anger	Iran		
Armyanskaya	Armenia		
Asin	Niger		
Asinara donkey (*Asino dell'Asinara*)	Italy	critically endangered; indigenous to the Isola dell'Asinara, Sardinia	
Asino Argentato di Sologno	Italy		
Asino Baio Lucano	Italy		
Asino dell'Irpinia	Italy	extinct; provinces of Benevento and Avellino	
Asino di Sant'Alberto	Italy	extinct; provinces of Forlì and Cesena	
Asino Emiliano	Italy	extinct	
Asino Pugliese	Italy	Puglia; regional variants include the Asino delle Marche, Asino della Basilicata and Asino Leccese	
Asino Sardo Grigio Crociato	Italy	from Sardinia	
Asnal Criolin'	Cuba		
Asno	Chile		
Asno Americana	Cuba		

Asno Balear	Spain		
Asno Criollo	Venezuela		
Asno de las Encartaciones	Spain	Basque Country	
Australian Donkey	Australia		
Azerbaidzhanskaya	Azerbaijan		
Balkan donkey (*Domaci balkanski magarac*)	Serbia, Montenegro		
Barockesel/Austrian-Hungarian white Donkey	Austria		
Benderi	Iran		
Berry Black (*Grand Noir du Berry*)	France	former province of Berry; possibly derived from Catalan	
Berry Grey (*Âne Petit Gris du Berry*)	France	extinct	
Biyang	China		
Black	Jordan		
Bourbonnais Donkey (*Âne du Bourbonnais*)	France		
Bourik	Haiti		
Brasil	Venezuela		
Bulgaro	Venezuela		

Burro	Mexico, Nicaragua, USA	small donkey of Mexico and the USA seen in both domesticated and feral states; some feral burros in the western United States are protected by federal law	
Burro Criollo	El Salvador		
Burro Kentucky	El Salvador, Honduras		
Caninde	Brazil		
Cardao	Brazil		
Cariovilli	Italy	extinct; province of L'Aquila in Abruzzo	
Castel Morrone donkey (*Asino di Castel Morrone*)	Italy	probably extinct; around Castel Morrone in the Province of Caserta, Campania	
Catalan donkey (*Catalana*)	Spain		
Chigetai	Kazakhstan		
China North	China		
Comune	Albania		
Cotentin Donkey (*Âne du Cotentin*)	France		
Creole	Guyana		
Criollo	Peru		
Cyprus donkey	Cyprus		
Dagestanskaya	Russian Federation		
Damascus	Israel, Syrian Arab Republic		
Dezhou	China		
Dongolawl	Sudan		
Donkey	Afghanistan		
Donkey	Ireland		
Donkey	Saint Kitts and Nevis		
Donkey	Uganda		
Dwarf donkey of Graciosa (*Burro de Graciosa*)	Portugal	Endangered, 90 left; also called Burro anão da Graciosa	
Egypt Baladi	Egypt		
Egyptian	Egypt		
English Donkey	Australia		

Etbai	Sudan		
Ezel	Netherlands		
Georgian Ass	Georgia		
Green	Barbados		
Grey	Suriname		
Grigio Siciliano (*Asino Grigio Siciliano*)	Italy	Sicily	
Guangling	China		
Guanzhong	China		
Hamadan	Iran, Russian Federation		
Hassawi	Egypt		
Huaibei	China		
Indian	India		
Iranian	Iran		
Irish Donkey	Australia		
Istarski magarac	Croatia	Istria	
Italian	Serbia		
Jack Norteamericano	Venezuela		
Jiami	China		
Jinnan	China		
Jirnrna	Ethiopia		
Kakhetinskaya	Russian Federation		
Kara-Kalpakskaya	Uzbekistan		
Karakaçan	Turkey		
Kashan	Iran		
Kassala	Eritrea		
Kazakhskaya	Kazakhstan		
Kirgizskaya	Kyrgyzstan		
Kulun	China		
Liangzhou	China		
Libyan	Libya		
Linxian	China		
Magyar parlagi szamár	Hungary		
Majorera	Spain		
Maltese donkey (*Hmar Malti*)	Malta		
Mannar	Sri Lanka		
Martina Franca donkey (*Asino di Martina Franca*)	Italy	Martina Franca and neighbouring areas in south-east Murgia, in the provinces of Bari, Taranto and Brindisi in Puglia	
Maryisltaya	Turkmenistan		
Masai	Kenya, Tanzania		
Masri	Egypt		
Merzifon	Turkey		
Meskhet-Dzhavakhetskaya	Uzbekistan, Turkmenistan		

Miniature	USA	Miniature Mediterranean Donkey	
Miniature Mediterranean Donkey	Italy	Sardinia & Sicily	
Miranda donkey (*Burro do Miranda*)	Portugal	Endangered status; also called Burro do Planalto Mirandes, Mirandes, Raça asinina de Miranda, or Transmontano	
Moldavian Local	Moldova		
Moroccan	Morocco		
Mula	Venezuela		
Muscat	Tanzania		
National Genatic	Yemen		
Native of North Africa	Mali, Mauritania, Senegal		
Nordestina	Brazil		
Norman donkey (*Âne Normand*)	France		
Ogaden	Ethiopia		
Pantesco (*Asino Pantesco* or *Asino di Pantelleria*)	Italy	from the island of Pantelleria, Sicily	
Paulista	Brazil		
Pega	Brazil		
Peruano	Venezuela		
Poitevin	France	see Poitou donkey or *Baudet de Poitou*	

Poitou donkey (*Baudet de Poitou*), Poitevin	France	the Poitou Donkey breed was developed for the sole purpose of producing mules. It is a large donkey breed with a very long shaggy coat and no dorsal stripe	
Ponui donkey	New Zealand		
Primorsko dinarski magarac	Croatia		
Provence donkey (*Âne de Provence*)	France	Provence	
Puttalam Buruwa	Sri Lanka		
Pyrenean donkey (*Âne des Pyrénées*)	France	Aquitaine, Midi Pyrénées, Languedoc Roussillon; also called Gascon; similar to Catalan but smaller	
Qinghai	China		
Qinqyang	China		
Qirmani	Yemen		
Qubressy	Jordan		
Ragusano donkey (*Asino Ragusano*)	Italy	from Ragusa in Sicily, and neighbouring areas	
Riffawi	Sudan		
Romagnolo donkey (*Asino Romagnolo*)	Italy	from Emilia–Romagna	
Romanian Donkey	Romania		
Saidi	Egypt		
Sardinian donkey (*Asino Sardo*)	Italy	from Sardinia	
Sennar	Ethiopia		
Shanbei	China		
Sibbianl	Yemen		
Sjeverno jadranski magarac	Croatia		

Somali	Djibouti, Kenya, Somalia, Yemen		
South-west	China		
Sperki breed	Pakistan		
Spotted	USA		
Standard	USA		
Subei	China		
Sudanese Pack	Sudan		
Syrian	Israel, Syrian Arab Republic		
Tadzhikskaya	Tajikistan		
Taihang	China		
Tibetan	China, Nepal		
Toposa	Sudan		
Tswana	Botswana		
Tunisian	Tunisia		
Turkmenskaya	Turkmenistan		
Uzbekskaya	Uzbekistan		
Viterbese (*Grigio Viterbese*)	Italy	province of Viterbo in Lazio	
Xinjiang	China		
Yangyuan	China		
Yunnan	China		
Zamorano-Leonés	Spain		
Âne Africain	Central African Republic, Chad		
Âne du Gourma	Mali		
Âne du Miankala	Mali		
Âne du plateau Dogon	Mali		
Âne du Sahel	Mall		
Âne du Yatenga	Mali		
Âne locale	Madagascar		
Âne wallon	Belgium		
Ânes	Togo		

REFERENCES

1. Breeds of Cattle – Oklahoma State University
2. Breeds of Cattle – Cattle.com
3. Breeds of Cattle – Cow World
4. Cattle Breeds – Embryoplus.com
5. Breeds of Cattle – Official 2nd Edition
6. Cattle Breeds of the World
7. Portuguese Cattle Breeds
8. EuReCa - Towards self-sustainable EUropean, REgional CAttle breeds

9. Ekarius, Carol (2007). Storey's Illustrated Guide to Poultry Breeds. 210 MAS MoCA Way, North Adams MA 01247: Storey Publishing. ISBN 978-1-58017-667-5.
10. Heinrichs, Christine (2007). How To Raise Chickens. Voyageur Press. ISBN 978-0-7938-0601-0.
11. Percy, Pam (2006). The Field Guide to Chickens. Suite 200, 380 Jackson St, St Paul MN 55101: Voyageur Press. ISBN 0-7603-2473-5.

CHAPTER 3

Methods of Animal Breeding

Chapter Contents

- **Introduction**
- **Crossbreeds In Specific Animals**
- **Inbreeding**
- **Mutation Breeding**
- **Outcrossing**
- **Preservation Breeding**
- **Selective Breeding**
- **Smart Breeding**
- **Hybrid (Biology)**
- **Hybrid Speciation**

INTRODUCTION

A crossbreed or crossbred usually refers to an organism with purebred parents of two different breeds, varieties, or populations. *Crossbreeding*, sometimes called "designer crossbreeding", refers to the process of breeding such an organism, often with the intention to create offspring that share the traits of both parent lineages, or producing an organism with hybrid vigor. While crossbreeding is used to maintain health and viability of organisms, irresponsible crossbreeding can also produce organisms of inferior quality or dilute a purebred gene pool to the point of extinction of a given breed of organism.

The term is also used at times to refer to a domestic animal of unknown ancestry where the breed status of only one parent or grandparent is known, though the term "mixed breed" is technically more accurate. The term outcross is used to describe a type of crossbreeding used within a purebred breed to increase the genetic diversity within the breed, particularly when there is a need to avoid inbreeding.

In animal breeding, *crossbreed* describes crosses within a single species, while *hybrid* refers to crosses between different species. In plant breeding terminology, the term *crossbreed* is uncommon. No universal term is used to distinguish hybridization or crossing within a population from those between populations, or even those between species.

A Labradoodle, a crossbreed between a poodle and a retriever

CROSSBREEDS IN SPECIFIC ANIMALS

Cats

The many newly developed and recognized breeds of domestic cat are crossbreeds between existing, well-established breeds (sometimes with limited hybridization with some wild species), to either combine selected traits from the foundation stock, or propagate a rare mutation without excessive inbreeding. However, some nascent breeds such as the Aegean cat are developed entirely from a local landrace population. Most experimental cat breeds are crossbreeds.

Cattle

In cattle, there are systems of crossbreeding. In many crossbreeds, one is larger than the other. One is used when the purebred females are particularly adapted to a specific environment, and are crossed with purebred bulls from another environment to produce a generation having traits of both parents.

Sheep

The large number of breeds of sheep, which vary greatly, creates an opportunity for crossbreeding to be used to tailor production of lambs to the goal of the individual stockman.

Llamas

Results of crossbreeding classic and woolly breeds of llama are unpredictable. The resulting offspring displays physical characteristics of either parent, or a mix of characteristics from both, periodically producing a fleeced llama. The results are increasingly unpredictable when both parents are crossbreeds, with possibility of the offspring displaying characteristics of a grandparent, not obvious in either parent.

Dogs

A *crossbred* dog is a cross between two (sometimes more) known breeds, and is usually distinguished from a *mixed-breed dog*, which has ancestry from many sources, some of which may not be known. Crossbreeds are popular, due to the belief that they have increased vigor without loss of attractiveness of the dog. Certain planned crossbreeding between purebred dogs of different breeds can produce puppies worth more than their purebred parents, due to a high demand.

Horses

The National Show Horse was developed from crossbreeding programs in the 1970s and 1980s that blended Arabian horse and American Saddlebred bloodlines

Crossbreeding in horses is often done with the intent of ultimately creating a new breed of horse. One type of modern crossbreeding in horses is used to create many of the warmblood breeds. Warmbloods are a type of horse used in the sport horse disciplines, usually registered in an open stud book by a studbook selection procedure that evaluates conformation, pedigree and, in some animals, a training or performance standard. Most warmblood breeds began as a cross of draft horse breeds on Thoroughbreds, but have, in some cases, developed over the past century to the point where they are considered to be a true-breeding population and have a closed stud book. Other types of recognized crossbreeding include that within the American Quarter Horse, which will register horses with one Thoroughbred parent and one registered Quarter Horse parent in the "Appendix" registry, and allow such animals full breed registration status as Quarter Horses if they meet a certain performance standard. Another well-known crossbred horse is the Anglo-Arabian, which may be produced by a purebred Arabian horse crossed on a Thoroughbred, or by various crosses of Anglo-Arabians with other Anglo-Arabians, as long as the ensuing animal never has more than 75% or less than 25% of each breed represented in its pedigree.

Hybrid Animals

A hybrid animal is one with parentage of two separate species, differentiating it from crossbred animals, which have parentage of the same species. Hybrids are usually, but not always, sterile.

One of the most ancient types of hybrid animal is the mule, a cross between a female horse and a male donkey. The liger is a hybrid cross between a male lion and female tiger. The yattle is a cross between a cow and a yak. Other crosses include the tigon (between a male tiger and female lion) and yakalo (between a yak and buffalo). The Incas recognized that hybrids of Lama glama (llama) and Lama pacos (alpaca) resulted in a hybrid with none of the advantages of either parent.

At one time it was thought that dogs and wolves were separate species, and the crosses between dogs and wolves were called wolf hybrids. Today wolves and dogs are both recognized as *Canis lupus*, but the old term "wolf hybrid" is still used.

Mixed Breeds

A mixed-breed animal is defined as having undocumented or unknown parentage, while a crossbreed generally has known, usually purebred parents of two distinct breeds or varieties. A dog of unknown parentage is often called a mixed-breed dog, "mutt" or "mongrel." A cat of unknown parentage is often referred to as domestic short-haired or domestic long-haired cat generically, and in some dialects is often called a "moggy". A horse of unknown bloodlines is a grade horse.

INBREEDING

Inbreeding is the production of offspring from the mating or breeding of individuals or organisms that are closely related genetically. By analogy, the term is used in human reproduction, but more commonly refers to the genetic disorders and other consequences that may arise from incestuous sexual relationships and consanguinity.

Inbreeding results in homozygosity, which can increase the chances of offspring being affected by recessive or deleterious traits. This generally leads to a decreased biological fitness of a population (called inbreeding depression), which is its ability to survive and reproduce. An individual who inherits such deleterious traits is referred to as *inbred*. The avoidance of expression of such deleterious recessive alleles caused by inbreeding, via inbreeding avoidance mechanisms, is the main selective reason for outcrossing. Crossbreeding between populations also often has positive effects on fitness-related traits.

Inbreeding is a technique used in selective breeding. In livestock breeding, breeders may use inbreeding when, for example, trying to establish a new and desirable trait in the stock, but will need to watch for undesirable characteristics in offspring, which can then be eliminated through further selective breeding or culling. Inbreeding is used to reveal deleterious recessive alleles, which can then be eliminated through assortative breeding or through culling. In plant breeding, inbred lines are used as stocks for the creation of hybrid lines to make use of the effects of heterosis. Inbreeding in plants also occurs naturally in the form of self-pollination.

Overview

Offspring of biologically related persons are subject to the possible effect of inbreeding, such as congenital birth defects. The chances of such disorders are increased when the biological parents are more closely related. (See coefficient of inbreeding.) This is because such pairings have a 25% probability of producing homozygous zygotes, resulting in offspring with two recessive alleles, which can produce disorders when these alleles are deleterious. (inbreeding depression.) Because most recessive alleles are rare in populations, it is unlikely that two unrelated marriage partners will both be carriers of the same deleterious allele. However, because close relatives share a large fraction of their alleles, the probability that any such deleterious allele is inherited from the common ancestor through both parents is increased dramatically. It should also be noted that, in terms of probability, for each homozygous recessive individual formed, there is an equal chance of producing a homozygous dominant individual, one completely devoid of the harmful allele. Contrary to common belief, inbreeding does not in itself alter allele frequencies, but rather increases the relative proportion of homozygotes to heterozygotes. However, because the increased proportion of deleterious homozygotes exposes the allele to natural selection, in the long run its frequency decreases more rapidly in inbred populations. In the short term, incestuous reproduction is expected to increase the number of spontaneous abortions of zygotes, perinatal deaths, and postnatal offspring with birth defects. The advantages of inbreeding may be the result of a tendency to preserve the structures of alleles interacting at different loci that have been adapted together by a common selective history.

Malformations or harmful traits can stay within a population due to a high homozygosity rate, and this will cause a population to become fixed for certain traits, like having too many bones in an area, like the vertebral column of wolves on Isle Royale or having cranial abnormalities, such as in Northern elephant seals, where their cranial bone length in the lower mandibular tooth row has changed. Having a high homozygosity rate is problematic for a population because it will unmask recessive deleterious alleles generated by mutations, reduce heterozygote advantage, and it is detrimental to the survival of small, endangered animal populations. When deleterious recessive alleles are unmasked due to the increased homozygosity generated by inbreeding, this can cause inbreeding depression. The authors think that it is possible that the severity of inbreeding depression can be diminished if natural selection can purge such alleles from populations during inbreeding. If inbreeding depression can be diminished by natural selection then some traits, harmful or not, can be reduced and change the future outlook on small, endangered populations.

There may also be other deleterious effects besides those caused by recessive diseases. Thus, similar immune systems may be more vulnerable to infectious diseases (Major histocompatibility complex and sexual selection).

Inbreeding history of the population should also be considered when discussing the variation in the severity of inbreeding depression between and within species. With persistent inbreeding, there is evidence that shows that inbreeding depression becoming less severe. This is associated with the unmasking and elimination of severely deleterious recessive alleles. However, inbreeding depression is not a temporary phenomenon because this elimination of deleterious recessive alleles will never be complete. Eliminating slightly deleterious mutations through inbreeding under moderate selection is not as effective. Fixation of alleles most likely occurs through Muller's ratchet, when an asexual population's genome accumulates deleterious mutations that are irreversible.

Despite all its disadvantages, inbreeding can also have a variety of advantages, such as reducing the cost of sex, reducing the recombination load, and allowing the expression of recessive advantageous phenotypes. It has been proposed that under circumstances when the advantages of inbreeding outweigh the disadvantages, preferential breeding within small groups could be promoted, potentially leading to speciation.

Genetic Disorders

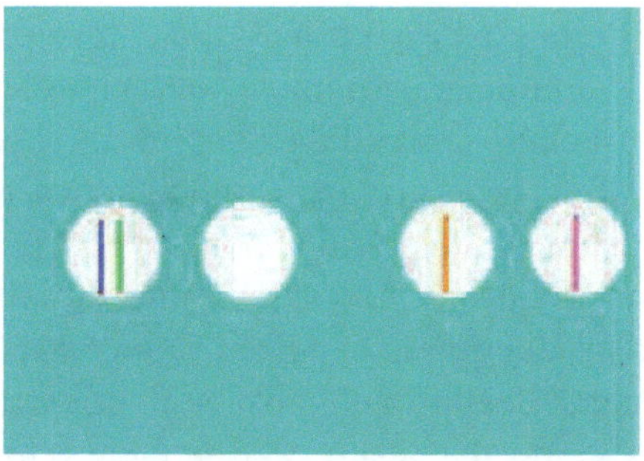

Animation of uniparental isodisomy

Autosomal recessive disorders occur in individuals who have two copies of an allele for a particular recessive genetic mutation. Except in certain rare circumstances, such as new mutations or uniparental disomy, both parents of an individual with such a disorder will be carriers of the gene. These carriers do not display any signs of the mutation and may be unaware that they carry the mutated gene. Since relatives share a higher proportion of their genes than do unrelated people, it is more likely that related parents will both be carriers of the same recessive allele, and therefore their children are at a higher risk of inheriting an autosomal recessive genetic disorder. The extent to which the risk increases depends on the degree of genetic relationship between the parents; the risk is greater when the parents are close relatives and lower for relationships between more distant relatives, such as second cousins, though still greater than for the general population.

Children of parent-child or sibling-sibling unions are at an increased risk compared to cousin-cousin unions. Inbreeding may result in a greater than expected phenotypic expression of deleterious recessive alleles within a population. As a result, first-generation inbred individuals are more likely to show physical and health defects, including:

- Reduced fertility both in litter size and sperm viability
- Increased genetic disorders
- Fluctuating facial asymmetry
- Lower birth rate
- Higher infant mortality and child mortality
- Smaller adult size
- Loss of immune system function
- Increased cardiovascular risks

The isolation of a small population for a period of time can lead to inbreeding within that population, resulting in increased genetic relatedness between breeding individuals. Inbreeding depression can also occur in a large population if individuals tend to mate with their relatives, instead of mating randomly.

Many individuals in the first generation of inbreeding will never live to reproduce. Over time, with isolation, such as a population bottleneck caused by purposeful (assortative) breeding or natural environmental factors, the deleterious inherited traits are culled.

Island species are often very inbred, as their isolation from the larger group on a mainland allows natural selection to work on their population. This type of isolation may result in the formation of race or even speciation, as the inbreeding first removes many deleterious genes, and permits the expression of genes that allow a population to adapt to an ecosystem. As the adaptation becomes more pronounced, the new species or race radiates from its entrance into the new space, or dies out if it cannot adapt and, most importantly, reproduce.

The reduced genetic diversity, for example due to a bottleneck will unavoidably increase inbreeding for the entire population. This may mean that a species may not be able to adapt to changes in environmental conditions. Each individual will have similar immune systems, as immune systems are genetically based. When a species becomes endangered, the population may fall below a minimum whereby the forced interbreeding between the remaining animals will result in extinction.

Natural breedings include inbreeding by necessity, and most animals only migrate when necessary. In many cases, the closest available mate is a mother, sister, grandmother, father, brother, or grandfather. In all cases, the environment presents stresses to remove from the population those individuals who cannot survive because of illness.

There was an assumption that wild populations do not inbreed; this is not what is observed in some cases in the wild. However, in species such as horses, animals in wild or feral conditions often drive off the young of both sexes, thought to be a mechanism by which the species instinctively avoids some of the genetic consequences of inbreeding. In general, many mammal species, including humanity's closest primate relatives, avoid close inbreeding possibly due to the deleterious effects.

Examples

Although there are several examples of inbred populations of wild animals, the negative consequences of this inbreeding are poorly documented.In the South American sea lion, there was concern that recent population crashes would reduce genetic diversity. Historical analysis indicated that a population expansion from just two matrilineal lines was responsible for most of the individuals within the population. Even so, the diversity within the lines allowed great variation in the gene pool that may help to protect the South American sea lion from extinction.

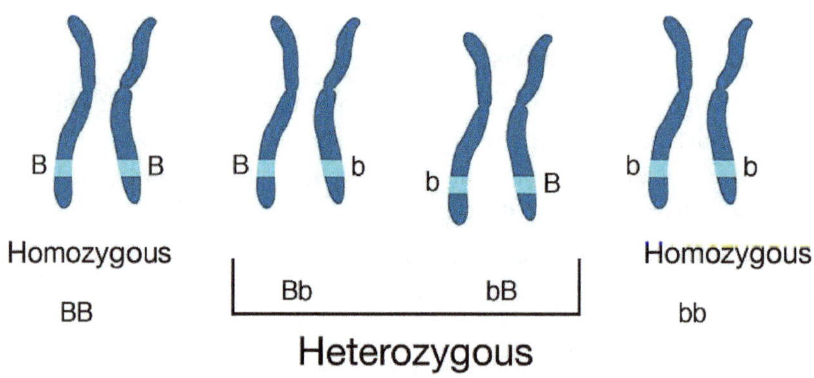

Heterozygous

In lions, prides are often followed by related males in bachelor groups. When the dominant male is killed or driven off by one of these bachelors, a father may be replaced by his son. There is no mechanism for preventing inbreeding or to ensure outcrossing. In the prides, most lionesses are related to one another. If there is more than one dominant male, the group of alpha males are usually related. Two lines are then being "line bred". Also, in some populations, such as the Crater lions, it is known that a population bottleneck has occurred. Researchers found far greater genetic heterozygosity than expected. In fact, predators are known for low genetic variance, along with most of the top portion of the trophic levels of an ecosystem. Additionally, the alpha males of two neighboring prides can be from the same litter; one brother may come to acquire leadership over another's pride, and subsequently mate with his 'nieces' or cousins. However, killing another male's cubs, upon the takeover, allows the new selected gene complement of the incoming alpha male to prevail over the previous male. There are genetic assays being scheduled for lions to determine their genetic diversity. The preliminary studies show results inconsistent with the outcrossing paradigm based on individual environments of the studied groups.

In Central California, Sea Otters were thought to have been driven to extinction due to over hunting, until a colony of about 30 breeding pairs was discovered in the Big Sur region in the 1930s. Since then, the population has grown and spread along the central Californian coast to around 2,000 individuals, a level that has remained stable for over a decade. Population growth is limited by the fact that all Californian Sea Otters are descended from the isolated colony, resulting in inbreeding.

Cheetahs are another example of inbreeding. Thousands of years ago the cheetah went through a population bottleneck that reduced its population dramatically so the animals that are alive today are all related to one another. A consequence from inbreeding for this species has been high juvenile mortality, low fecundity, and poor breeding success.

In a study on an island population of song sparrows, individuals that were inbred showed significantly lower survival rates than outbred individuals during a severe winter weather related population crash. These studies show that inbreeding depression and ecological factors have an influence on survival.

Measures of Inbreeding

A measure of inbreeding of an individual A is the probability *F*(A) that both alleles in one locus are derived from the same allele in an ancestor. These two identical alleles that are both derived from a common ancestor are said to be identical by descent. This probability F(A) is called the "coefficient of inbreeding".

Another useful measure that describes the extent to which two individuals are related (say individuals A and B) is their coancestry coefficient f(A,B), which gives the probability that one randomly selected allele from A and another randomly selected allele from B are identical by descent. This is also denoted as the kinship coefficient between A and B.

A particular case is the self-coancestry of individual A with itself, f(A,A), which is the probability that taking one random allele from A and then, independently and with replacement, another random allele also from A, both are identical by descent. Since they can be identical by descent by sampling the same allele or by sampling both alleles that happen to be identical by descent, we have $f(A,A) = 1/2 + F(A)/2$.

Both the inbreeding and the coancestry coefficients can be defined for specific individuals or as average population values. They can be computed from genealogies or estimated from the population size and its breeding properties, but all methods assume no selection and are limited to neutral alleles.

There are several methods to compute this percentage. The two main ways are the path method and the tabular method.

Typical coancestries between relatives are as follows:

- Father/daughter, mother/son or brother/sister → 25% ($^1/_4$)

- Grandfather/granddaughter or grandmother/grandson → 12.5% ($^1/_8$)
- Half-brother/half-sister, Double cousins → 12.5% ($^1/_8$)
- Uncle/niece or aunt/nephew → 12.5% ($^1/_8$)
- Great-grandfather/great-granddaughter or great-grandmother/great-grandson → 6.25% ($^1/_{16}$)
- Half-uncle/niece or half-aunt/nephew → 6.25% ($^1/_{16}$)
- First cousins → 6.25% ($^1/_{16}$)

Domestic Animals

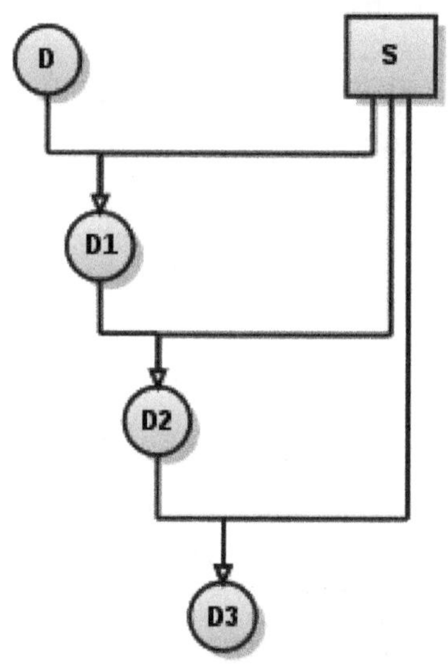

An intensive form of inbreeding where an individual **S** is mated to his daughter **D1**, granddaughter **D2** and so on, in order to maximise the percentage of **S**›s genes in the offspring. 87.5% of **D3**'s genes would come from **S**, while **D4** would receive 93.75% of their genes from **S**.

Breeding in domestic animals is primarily assortative breeding (see selective breeding). Without the sorting of individuals by trait, a breed could not be established, nor could poor genetic material be removed. Homozygosity is the case where similar or identical alleles combine to express a trait that is not otherwise expressed (recessiveness). Inbreeding exposes recessive alleles through increasing homozygosity.

Breeders must avoid breeding from individuals that demonstrate either homozygosity or heterozygosity for disease causing alleles. The goal of preventing the transfer of deleterious alleles may be achieved by reproductive isolation, sterilization, or, in the extreme case, culling. Culling is not strictly necessary if genetics are the only issue in hand. Small animals such as cats and dogs may be sterilized, but in the case of large agricultural animals, such as cattle, culling is usually the only economic option.

The issue of casual breeders who inbreed irresponsibly is discussed in the following quotation on cattle:

Meanwhile, milk production per cow per lactation increased from 17,444 lbs to 25,013 lbs from 1978 to 1998 for the Holstein breed. Mean breeding values for milk of Holstein cows increased by 4,829 lbs during this period. High producing cows are increasingly difficult to breed and are subject to higher health costs than cows of lower genetic merit for production (Cassell, 2001).

Intensive selection for higher yield has increased relationships among animals within breed and increased the rate of casual inbreeding.

Many of the traits that affect profitability in crosses of modern dairy breeds have not been studied in designed experiments. Indeed, all crossbreeding research involving North American breeds and strains is very dated (McAllister, 2001) if it exists at all.

The BBC produced two documentaries on dog inbreeding titled Pedigree Dogs Exposed and Pedigree Dogs Exposed - Three Years On that document the negative health consequences of excessive inbreeding.

Linebreeding is a form of inbreeding. There is no clear distinction between the two terms, but linebreeding may encompass crosses between individuals and their descendants or two cousins. This method can be used to increase a particular animal's contribution to the population. While linebreeding is less likely to cause problems in the first generation than does inbreeding, over time, linebreeding can reduce the genetic diversity of a population and cause problems related to a too-small genepool that may include an increased prevalence of genetic disorders and inbreeding depression.

Outcrossing is where two unrelated individuals are crossed to produce progeny. In outcrossing, unless there is verifiable genetic information, one may find that all individuals are distantly related to an ancient progenitor. If the trait carries throughout a population, all individuals can have this trait. This is called the founder effect. In the well established breeds, that are commonly bred, a large gene pool is present. For example, in 2004, over 18,000 Persian cats were registered. A possibility exists for a complete outcross, if no barriers exist between the individuals to breed. However, it is not always the case, and a form of distant linebreeding occurs. Again it is up to the assortative breeder to know what sort of traits, both positive and negative, exist within the diversity of one breeding. This diversity of genetic expression, within even close relatives, increases the variability and diversity of viable stock.

Laboratory Animals

Systematic inbreeding and maintenance of inbred strains of laboratory mice and rats is of great importance for biomedical research. The inbreeding guarantees a consistent and uniform animal model for experimental purposes and enables genetic studies in congenic and knock-out animals. The use of inbred strains is also important for genetic studies in animal models, for example to distinguish genetic from environmental effects. The mice that are inbred typically show considerably lower survival rates.

Humans

Effects

Inbreeding increases the chances of the expression of deleterious recessive alleles by increasing homozygosity and therefore has the potential to decrease the fitness of the offspring. With continuous inbreeding, genetic variation is lost and homozygosity is increased, enabling the expression of recessive deleterious alleles in homozygotes. The inbreeding coefficient, a term used to describe the degree of inbreeding in an individual, is an estimate of the percent of homozygous alleles in the overall genome. The more biologically related the parents are, the greater the inbreeding coefficient (Coefficient of Inbreeding), since their genomes have many similarities already. This overall homozygosity becomes an issue when there are deleterious recessive alleles in the gene pool of the family. By pairing chromosomes of similar genomes, the chance for these recessive alleles to pair and become homozygous greatly increases, leading to offspring with autosomal recessive disorders.

Inbreeding is especially problematic in small populations where the genetic variation is already limited. By inbreeding, individuals are further decreasing genetic variation by increasing homozygosity in the genomes of their offspring. Thus, the likelihood of deleterious recessive alleles to pair is significantly higher in a small inbreeding population than in a larger inbreeding population.

The fitness consequences of consanguineous mating have been studied since their scientific recognition by Charles Darwin in 1839. Some of the most harmful effects known from such breeding includes its effects on the mortality rate as well as on the general health of the offspring. Within the past several decades, there have been many studies to support such debilitating effects on the human organism. Specifically, inbreeding has been found to decrease fertility as a direct result of increasing homozygosity of deleterious recessive alleles. Fetuses produced by inbreeding also face a greater risk of spontaneous abortions due to inherent complications in development. Among mothers who experience stillbirths and early infant deaths, those that are inbreeding have a significantly higher chance of reaching repeated results with future offspring. Additionally, consanguineous parents possess a high risk of premature birth and producing underweight and undersized infants. Viable inbred offspring are also likely to be inflicted with physical deformities and genetically inherited diseases. Studies have confirmed an increase in several genetic disorders due to inbreeding such as blindness, hearing loss, neonatal diabetes, limb malformations, Schizophrenia and several others. Moreover, there is an increased risk for congenital heart disease depending on the inbreeding coefficient (Coefficient of Inbreeding) of the offspring, with significant risk accompanied by an F =.125 or higher.

Prevalence

The general negative outlook and eschewal of inbreeding that is prevalent in the Western world today holds roots from over 1500 years ago. Specifically, written documents such as the Bible illustrate that there have been laws and social customs that have called for the abstention from inbreeding. Along with cultural taboos, parental education and awareness of inbreeding consequences have played large roles in minimizing inbreeding frequencies in areas like Europe. That being so, there are less urbanized and less populated regions across the world that have shown continuity in the practice of inbreeding. This continuity is often either by choice or unavoidably due to the limitations of the geographical area. When by choice, the rate of consanguinity is highly dependent on religion and culture. Of the practicing regions, Middle Eastern and northern Africa territories show the greatest frequencies of consanguinity. The link between the high frequency and the region is primarily due to the dominance of Islamic populations, who have historically engaged in familyline relations.

- Qatar: Historically, populations of Qatar have engaged in consanguineous relationships of all kinds, leading to high risk of inheriting genetic diseases. As of 2014, around 5% of the Qatari population suffered from hereditary hearing loss; most were descendants of a consanguineous relationship.

Among these populations with high levels of inbreeding, researchers have found several disorders prevalent among inbred offspring. Specifically, in Lebanon, Saudi Arabia, Egypt, and Arabs in Israel, it has been discovered that offspring of consanguineous relationships have an increased risk of congenital malformations, congenital heart defects, congenital hydrocephalus and neural tube defects. Furthermore, among inbred children in Palestine and Lebanon, there is a positive association between consanguinity and reported cleft lip/palate cases.

Royalty and Nobility

Inter-nobility marriage was used as a method of forming political alliances among elites. These ties were often sealed only upon the birth of progeny within the arranged marriage. Thus marriage was seen as a union of lines of nobility, not as a contract between individuals as it is seen today.

Royal intermarriage was often practiced among European royal families, usually for interests of state. Over time, due to the relatively limited number of potential consorts, the gene pool of many ruling families grew progressively smaller, until all European royalty was related. This also resulted in many being descended from a certain person through many lines of descent, such as the numerous European royalty and nobility descended from the British Queen Victoria or King Christian IX of Denmark. The House of Habsburg was infamous for its inbreeding, with the Habsburg lip cited as an ill-effect, although no genetic evidence has proved the allegation. The closely related houses of Habsburg, Bourbon, Braganza and Wittelsbach also frequently engaged in first-cousin unions as well as the occasional double-cousin and uncle-niece marriages. Examples of incestuous marriages and the impact of inbreeding on royal families include:

- In ancient Egypt, royal women were believed to carry the bloodlines and so it was advantageous for a pharaoh to marry his sister or half-sister; in such cases a special combination between endogamy and polygamy is found. Normally, the old ruler's eldest son and daughter (who could be either siblings or half-siblings) became the new rulers. All rulers of the Ptolemaic dynasty uninterruptedly from Ptolemy IV (Ptolemy II married his sister but had no issue) were married to their brothers and sisters, so as to keep the Ptolemaic blood "pure" and to strengthen the line of succession. Cleopatra VII (also called Cleopatra VI) and Ptolemy XIII, who married and became co-rulers of ancient Egypt following their father's death, are the most widely known example.

- One of the most famous examples of a genetic trait aggravated by royal family intermarriage was the House of Habsburg, which inmarried particularly often and is known for the mandibular prognathism of the *Habsburger (Unter) Lippe* (otherwise known as the 'Habsburg jaw', 'Habsburg lip' or 'Austrian lip'"). This was typical for many Habsburg relatives over a period of six centuries. The condition progressed through the generations to the point that the last of the Spanish Habsburgs, Charles II of Spain, could not properly chew his food. Besides the jaw deformity, Charles II also had a huge number of other genetic physical, intellectual, sexual, and emotional problems. It is speculated that the simultaneous occurrence in Charles II of two different genetic disorders (combined pituitary hormone deficiency and distal renal tubular acidosis) could explain most of the complex clinical profile of this king, including his impotence/infertility, which led to the extinction of the dynasty.

MUTATION BREEDING

Mutation breeding, sometimes referred to as "variation breeding", is the process of exposing seeds to chemicals or radiation in order to generate mutants with desirable traits to be bred with other cultivars. Plants created using mutagenesis are sometimes called mutagenic plants or mutagenic seeds. From 1930 to 2014 more than 3200 mutagenic plant varietals have been released that have been derived either as direct mutants (70%) or from their progeny (30%). Crop plants account for 75% of released mutagenic species with the remaining 25% ornamentals or decorative plants. However, although the FAO/ IAEA reported in 2014 that over 1,000 mutant varietals of major staple crops were being grown worldwide, it is unclear how many of these varieties are currently used in agriculture or horticulture around the world, as these seeds are not always identified or labeled as being mutagenic or having a mutagenic provenance.

Process

There are different kinds of mutagenic breeding such as using chemical mutagens like ethyl methanesulfonate and dimethyl sulfate, radiation and transposons are used to generate mutants. Mutation breeding is commonly used to produce traits in crops such as larger seeds, new colors, or sweeter fruits, that either cannot be found in nature or have been lost during evolution.

Radiation Breeding

Exposing plants to radiation is sometimes called radiation breeding and is a sub class of mutagenic breeding. Radiation breeding was discovered in the 1920s when Lewis Stadler of the University of Missouri used X-rays on maize and barley. In the case of barley, the resulting plants were white, yellow, pale yellow and some had white stripes. In 1928, Stadler first published his findings on radiation-induced mutagenesis in plants. During the period 1930-2004, radiation-induced mutant varieties were developed primarily using gamma rays (64%) and X-rays (22%).[:187]

Radiation breeding may take place in atomic gardens; and seeds have been sent into orbit in order to expose them to more cosmic radiation.

Use of Chemical Mutagens

High rates of chromosome aberrations resulting from ionizing radiation and the accompanied detrimental effects made researchers look for alternate sources for inducing mutations. As a result, an array of chemical mutagens has been discovered. The most widely used chemical mutagens are alkylating agents. Ethyl methanesulfonate (EMS) is the most popular because of its effectiveness and ease of handling, especially its detoxification through hydrolysis for disposal. Nitroso compounds are the other alkylating agents widely used, but they are light-sensitive and more precautions need to be taken because of their higher volatility. EMS has become a commonly used mutagen for developing large numbers of mutants for screening such as in developing TILLING populations. Although many chemicals are mutagenic, only few have been used in practical breeding as the doses need to optimised and also because the effectiveness is not high in plants for many.

History

After WWII, there was a concerted effort to find 'peaceful' uses for atomic energy. One of the ideas was to bombard plants with radiation and produce lots of mutations, some of which, it was hoped, would lead to plants that bore more heavily or were disease or cold-resistant or just had unusual colors. The experiments were mostly conducted in giant gamma gardens on the grounds of national laboratories in the US but also in Europe and countries of the former USSR.

Comparison to Other Agronomic Techniques

In the debate over Genetically Modified foods, the use of transgenic processes is often compared and contrasted with mutagenic processes. While the abundance and variation of transgenic organisms in human food systems, and their effect on agricultural biodiversity, ecosystem health and human health is somewhat well documented, mutagenic plants and their role on human food systems is less well known, with one journalist writing "Though poorly known, radiation breeding has produced thousands of useful mutants and a sizable fraction of the world's crops, including varieties of rice, wheat, barley, pears, peas, cotton, peppermint, sunflowers, peanuts, grapefruit, sesame, bananas, cassava and sorghum." In Canada crops generated by mutation breeding face the same regulations and testing as crops obtained by genetic engineering. Mutagenic varieties tend to be made freely available for plant breeding, in contrast to many commercial plant varieties or germplasm that increasingly have restrictions on their use such as terms of use, patents and proposed genetic user restriction technologies and other intellectual property regimes and modes of enforcement.

Unlike genetically modified crops, which typically involve the insertion of one or two target genes, plants developed via mutagenic processes with random, multiple and unspecific genetic changes have been discussed as a concern but are not prohibited by any nation›s organic standards. Reports from the US National Academy of Sciences state that there is no scientific justification for regulating genetic engineered crops while not doing so for mutation breeding crops.

Several organic food and seed companies promote and sell certified organic products that were developed using both chemical and nuclear mutagenesis.Several certified organic brands, whose companies support strict labeling or outright bans on GMO-crops, market their use of branded wheat and other varietal strains which were derived from mutagenic processes without any reference to this genetic manipulation. These organic products range from mutagenic barley and wheat ingredient used in organic beers to mutagenic varieties of grapefruits sold directly to consumers as organic.

New Mutagenic Techniques

Restriction Endonucleases

Interest in the use of bacterial restriction endonucleases (RE) to study double-stranded breaks in plant DNA began in the mid-nineties. These breaks in DNA, otherwise known as DSBs, were found to be the source of much chromosomal damage in eukaryotes, causing mutations in plant varietals. REs induce a result on plant DNA similar to that of ionizing radiation or radiomimetic chemicals. Blunt ended breaks in the DNA, unlike sticky ended breaks, were found to produce more variations in chromosomal damage, making them the more useful type of break for mutation breeding. While the connection of REs to chromosomal aberrations is mostly limited to research on mammalian DNA, success in mammalian studies caused scientists to conduct more studies of RE-induced chromosomal and DNA damaged on barley genomes. Due to restriction endonucleases' ability to facilitate damage in chromosomes and DNA, REs have the capability of being used as a new method of mutagenesis to promote the proliferation of mutated plant varietals.

Space-Breeding

The ability of plants to develop and thrive is dependent on conditions such as microgravity and cosmic radiation in space. China has been experimenting with this theory by sending seeds into space, testing to see if space flights will cause genetic mutations. Since 1987, China has cultivated 66 mutant varieties from space through their space-breeding program. Chromosomal aberrations greatly increased when seeds were sent into aerospace compared to their earth-bound counterparts. The effect of space flight on seeds depends on their species and variety. For example, space-bred wheat saw a large growth in seed germination in compared to its Earth-bound control, but space-bred rice had no visible advantage compared to its control. For the varietals that were positively mutated by space flight, their growth potential exceeded that of not only their Earth-grown counterparts, but also their irradiated counterparts on Earth. Compared to traditional mutagenic techniques, space-bred mutations have greater efficacy in that they experience positive effects on their first generation of mutation, whereas irradiated crops often see no advantageous mutations in their first generations. Though multiple experiments have shown the positive effects of space flight on seed mutation, there is no clear connection as to what aspect of aerospace has produced such advantageous mutations. There is much speculation around cosmic radiation being the source of chromosomal aberrations, but so far, there has been no concrete evidence of such connection. Though China's space-breeding program has been shown to be very successful, the program requires a large budget and technological support that many other countries are either unwilling or unable to provide, meaning this program is unfeasible outside of China. Due to such restraints, scientists have been trying to replicate space condition on Earth in order to promote the same expedient space-born mutations on Earth. One such replication is a magnetic field-free space (MF), which produces an area with a weaker magnetic field than that of Earth. MF treatment produced mutagenic results, and has been used to cultivate new mutant varieties of rice and alfalfa. Other replications of space conditions include irradiation of seeds by a heavy 7Li ion beam or mixed high-energy particles. These space-bred varietals are already being introduced to the public. In 2011, during the National Lotus Flowers Exhibition in China, a mutant lotus, called the "Outer Space Sun", was shown at the flower show.

Ion Beam Technology

Ion beams mutate DNA by deleting multiple bases from its code. Compared to traditional sources of radiation, like Gamma rays and X-rays, ion beams have been shown to cause more severe breaks in DNA that are more difficult to weave back together, causing the change in DNA to be more drastic than changes caused by traditional irradiation. Ion beams change DNA in a manner that makes it look vastly different than its original makeup, more so than when traditional irradiation techniques are used. Most experimentation, using ion beam technology, has been conducted in Japan. Notable facilities using this technology are TIARA of the Japan Atomic Energy Agency, RIKEN Accelerator Research Facility, and various other Japanese institutions. During the process of ion beam radiation, seeds are wedged between two kapton films and irradiated for roughly two minutes. Mutation frequencies are notably higher for ion beam radiation compared to electron radiation, and the mutation spectrum is broader for ion beam radiation compared to Gamma ray radiation. The broader mutation spectrum was revealed through the largely varied amount of flower phenotypes produced by ion beams. Flowers mutated by the ion beams exhibited a variety of colors, patterns, and shapes. Through ion beam radiation, new varieties of plants have been cultivated. These plants had the characteristics of being ultraviolet light-B resistant, disease resistant, and chlorophyll-deficient. Ion beam technology has been used in the discovery of new genes responsible for the creation of more robust plants, but its most prevalent use is commercially for producing new flower phenotypes, like striped chrysanthemums.

Mature Pollen Treated with Gamma Radiation

Gamma radiation is used on mature rice pollen to produce parent plants used for crossing. The mutated traits in the parent plants are able to be inherited by their offspring plants. Because rice pollen has a very short lifespan, researchers had to blast Gamma rays at cultured spikes from rice plants. Through experimentation, it was revealed that there was a greater variety of mutation in irradiated pollen rather than irradiated dry seeds. Pollen treated with 46Gy of gamma radiation showed an increase in grain size overall and other useful variations. Typically, the length of each grain was longer after the crossing of irradiated parent rice plants. The rice progeny also exhibited a less chalky visage, improving on the appearance of the parent rice plants. This technique was used to develop two new rice cultivars, Jiaohezaozhan and Jiafuzhan, in China. Along with facilitating the creation of these two rice cultivars, the irradiation of mature rice pollen has produced roughly two hundred mutant rice lines. Each of these lines produce rice grains of both a higher quality and larger size. The mutations produced by this technique vary with each generation, meaning further breeding of these mutated plants could produce new mutations. Traditionally, gamma radiation is used on solely adult plants, and not on pollen. The irradiation of mature pollen allows mutant plants to grow without being in direct contact with gamma radiation. This discovery is in contrast to what was previously believed about gamma radiation: that it could only elicit mutations in plants and not pollen.

Notable Mutagenized Varietals

Argentina

- Colorado Irradiado groundnut (mutant created with X-rays; high fat content and yield, 80% of groundnuts grown in Argentina the 1980s was Colorado Irradiado)
- Puita INTA-CL rice mutant (herbicide resistance and good yield; also grown in Bolivia, Brazil, Costa Rica and Paraguay)

Australia

- Amaroo rice mutant variety (60-70% of rice grown in Australia was Amaroo in 2001)

Bangladesh

- Binasail, Iratom-24 and Binadhan-6 rice mutants
- Binamoog-5 mung bean mutant variety

Cuba

- Maybel tomato mutant (excellent drought resistance)
- GINES rice mutant (created using proton radiation; grows well in salty conditions)

People's Republic of China

- Henong series soybean mutants
- Jiahezazhan and Jiafuzhan rice (mutations obtained by pollen irradiation; high

yield and quality, very adaptable, resistant to plant hopper and blast)

- Lumian Number 1 cotton
- Purple Orchard 3 Sweet potato
- Tiefeng 18 soybean
- Yangdao Number 6 rice
- Yangmai 156 wheat
- Zhefu 802 rice mutant (irradiated with gamma rays; resistant to rice blast, good yield even in poor conditions, the most planted rice vatriety between 1986-1994)
- 26Zhaizao indica rice mutant (created with gamma rays)

Czech Republic

- Diamant barley (high yield, short height mutant created with X-Rays)

Egypt

- Giza 176 and Sakha 101 high yield rice mutants

Finland

- Balder J barley mutant (better drought resistance, yield and sprouting)
- Puhti and Ryhti stiff straw oat mutants

Germany

- Trumpf barley

Ghana

- Tek bankye mutant cassava (good poundability and increased dry matter content)

India

- Co-4, Pant Mung-2, and TAP mung bean mutants
- MA-9 cotton - the world›s first mutant cotton, released in 1948 (X-ray radiation; drought tolerance)
- PNR-381 Rice
- Pusa 408 (Ajay), Pusa 413 (Atul), Pusa 417 (Girnar), and Pusa – 547 chickpea mutants (resistant to Ascochyta blight and wilt diseases, and have high yields)
- Sharbati Sonora wheat
- Tau-1, MUM 2, BM 4, LGG 407, LGG 450, Co4, Dhauli (TT9E) and Pant moong-1 blackgram (YMC, (Yellow mosaic virus) resistance)
- TG24 and TG37 groundnut mutants

Italy

- Durum wheat (especially Creso mutant, created with thermal neutrons)

Japan

- Osa Gold Pear (disease resistance)
- Most rice varieties grown in Japan have the sd1 mutant allele from the Reimei rice variety

Myanmar

- Shwewartun rice mutant (created by irradiating IR5 rice to give better yield, grain quality and earlier maturity)

Pakistan

- Basmati 370 short height rice mutant
- NIAB-78 cotton mutant (high yielding, heat tolerant, early maturing)
- CM-72 chickpea mutant (created with 150 Gy of gamma rays; high yielding, blight resistant, desi type)
- NM-28 mungbean mutant (short height, uniform and ealry maturing, high seed yield)
- NIAB Masoor 2006 lentil mutant (created with 200 Gy of radiation; early maturing, high yield, resistant to disease)

Peru

- UNA La Molina 95 barley mutant (developed in 1995 for growing above 3,000 m)
- Centenario Amarinth «kiwicha» mutant (high quality grain and exported as a certified organic product)
- Centenario II barley mutant (developed for growing in the Andean highlands with high yield, high quality flour and tolerance to hail)

Sudan

- Albeely banana mutant (better quality, high yield and better stand)

Thailand

- RD15 and RD6 aromatic indica rice mutants (created with gamma rays and released in 1977-8; RD 15 is early ripening, RD6 has a valuable glutinous endosperm) Thailand is the biggest exporter of aromatic rice in the world

United Kingdom

- Golden Promise barley (semi-dwarf, salt tolerant mutant created with gamma rays) Is used to make beer and whisky

United States

- Calrose 76 Rice (short height rice induced with gamma rays)[189]
- Luther and Pennrad barley (high yield mutant varities; Pennrad also resistant to winter)
- Murray Mitcham Peppermint (*Verticillium* wilt tolerance)[189]
- Sanilac bean (X-ray radiation; high yielding mutant - also the Gratiot and Sea-

way bean varieties were cross-bred from Sanilac)

- Stadler wheat (high yield mutant with resistance to loose smut and leaf rust and earlier maturity)
- Star Ruby and Rio red varieties of the Rio Star Grapefruit (created using thermal neutron techniques)[:189]
- Todd's Mitcham Peppermint (*Verticillium* wilt tolerance)[:189]

Vietnam

- VND 95-20, VND-99-1 and VN121 rice mutants (increased yield, improved quality, resistance to disease and pests)
- DT84, DT96, DT99 and DT 2008 soybean mutants (developed using gamma rays to grow three crops a year, tolerance to heat and cold and resistance to disease)

In 2014, it was reported that 17 rice mutant varietals, 10 soybean, two maize and one chrysanthemum mutant varietals had been officially released to Vietnamese farmers. 15% of rice and 50% of soybean was produced from mutant varietals.

OUTCROSSING

Outcrossing or **outbreeding** is the practice of introducing unrelated genetic material into a breeding line. It increases genetic diversity, thus reducing the probability of an individual being subject to disease or genetic abnormalities.

Outcrossing is now the norm of most purposeful animal breeding, contrary to what is commonly believed. The outcrossing breeder intends to remove the traits by using «new blood». With dominant traits, one can still see the expression of the traits and can remove those traits whether one outcrosses, line breeds or inbreds. With recessive traits, outcrossing allows for the recessive traits to migrate across a population. The outcrossing breeder then may have individuals that have many deleterious genes that may be expressed by subsequent inbreeding. There is now a gamut of deleterious genes within each individual in many dog breeds.

Increasing the variation of genes or alleles within the gene pool may protect against extinction by stressors from the environment. For example, in this context, a recent veterinary medicine study tried to determine the genetic diversity within cat breeds.

Outcrossing is believed to be the «norm» in the wild. Outcrossing in plants is usually enforced by self-incompatibility.

Breeders inbreed within their genetic pool, attempting to maintain desirable traits and to cull those traits that are undesirable. When undesirable traits begin to appear, mates are selected to determine if a trait is recessive or dominant. Removal of the trait is accomplished by breeding two individuals known not to carry it.

Gregor Mendel used outcrossing in his experiments with flowers. He then used the resulting offspring to chart inheritance patterns, using the crossing of siblings, and backcrossing to parents to determine how inheritance functioned.

Charles Darwin, in his book "The Effects of Cross and Self-Fertilization in the Vegetable Kingdom," came to clear and definite conclusions concerning the adaptive benefit of outcrossing. For example, he stated (on page 462) that "the offspring from the union of two distinct individuals, especially if their progenitors have been subjected to very different conditions, have an immense advantage in height, weight, constitutional vigor and fertility over the self-fertilizing offspring from either one of the same parents." He thought that this observation was amply sufficient to account for outcrossing sexual reproduction. The disadvantages of self-fertilized offspring (inbreeding depression) are now thought to be largely due to the homozygous expression of deleterious recessive mutations; and the fitness advantages of outcrossed offspring are thought to be largely due to the heterozygous masking of such deleterious mutations.

PRESERVATION BREEDING

Preservation breeding is an attempt by many animal breeders to preserve bloodlines of animals, either of a rare breed, or of rare pedigrees within a breed. One purpose of preservation breeding is to protect genetic diversity within a species, another is to preserve valuable genetic traits that may not be popular or in fashion in the present, but may be of great value in the future. Preservation breeding can also mean the breeding of animals to populate or repopulate an area where a species previously existed. It can also mean the breeding of species that are healthy and releasing the healthy individuals into a population that is defective or infected for the purpose of strengthening the overall health of the population.

SELECTIVE BREEDING

Selective breeding (also called artificial selection) is the process by which humans use animal breeding and plant breeding to selectively develop particular phenotypic traits (characteristics) by choosing which typically animal or plant males and females will sexually reproduce and have offspring together. Domesticated animals are known as breeds, normally bred by a professional breeder, while domesticated plants are known as varieties, cultigens, or cultivars. Two purebred animals of different breeds produce a crossbreed, and crossbred plants are called hybrids. Flowers, vegetables and fruit-trees may be bred by amateurs and commercial or non-commercial professionals: major crops are usually the provenance of the professionals.

There are two approaches or types of artificial selection, or selective breeding. First is the traditional «breeder's approach» in which the breeder or experimenter applies «a known amount of selection to a single phenotypic trait» by examining the chosen trait and choosing to breed only those that exhibit higher or «extreme values» of that trait. The second is called «controlled natural selection,» which is essentially natural selection in a controlled environment. In this, the breeder does not choose which individuals being tested «survive or reproduce,» as he or she could in the traditional approach. There are also «selection experiments,» which is a third approach and these are conducted in order to determine the «strength of natural selection in the wild.» However, this is more often an observational approach as opposed to an experimental approach.

In animal breeding, techniques such as inbreeding, linebreeding, and outcrossing are utilized. In plant breeding, similar methods are used. Charles Darwin discussed how selective breeding had been successful in producing change over time in his 1859 book, *On the Origin of Species.* Its first chapter discusses selective breeding and domestication of such animals as pigeons, cats, cattle, and dogs. Darwin used artificial selection as a springboard to introduce and support the theory of natural selection.

The deliberate exploitation of selective breeding to produce desired results has become very common in agriculture and experimental biology.

Selective breeding can be unintentional, e.g., resulting from the process of human cultivation; and it may also produce unintended – desirable or undesirable – results. For example, in some grains, an increase in seed size may have resulted from certain ploughing practices rather than from the intentional selection of larger seeds. Most likely, there has been an interdependence between natural and artificial factors that have resulted in plant domestication.

A Belgian Blue cow. The defect in the breed's myostatin gene is maintained through linebreeding and is responsible for its accelerated lean muscle growth.

This Chihuahua mix and Great Dane shows the wide range of dog breed sizes created using selective breeding.

History

Selective breeding of both plants and animals has been practiced since early prehistory; key species such as wheat, rice, and dogs have been significantly different from their wild ancestors for millennia, and maize, which required especially large changes from teosinte, its wild form, was selectively bred in Mesoamerica. Selective breeding was practiced by the Romans. Treatises as much as 2,000 years old give advice on selecting animals for different purposes, and these ancient works cite still older authorities, such as Mago the Carthaginian. The notion of selective breeding was later expressed by the Persian Muslim polymath Abu Rayhan Biruni in the 11th century. He noted the idea in his book titled *India*, which included various examples.

The agriculturist selects his corn, letting grow as much as he requires, and tearing out the remainder. The forester leaves those branches which he perceives to be excellent, whilst he cuts away all others. The bees kill those of their kind who only eat, but do not work in their beehive.

—Abu Rayhan Biruni, India

Selective breeding was established as a scientific practice by Robert Bakewell during the British Agricultural Revolution in the 18th century. Arguably, his most important breeding program was with sheep. Using native stock, he was able to quickly select for large, yet fine-boned sheep, with long, lustrous wool. The Lincoln Longwool was improved by Bakewell, and in turn the Lincoln was used to develop the subsequent breed, named the New (or Dishley) Leicester. It was hornless and had a square, meaty body with straight top lines.

These sheep were exported widely, including to Australia and North America, and have contributed to numerous modern breeds, despite the fact that they fell quickly out of favor as market preferences in meat and textiles changed. Bloodlines of these original New Leicesters survive today as the English Leicester (or Leicester Longwool), which is primarily kept for wool production.

Bakewell was also the first to breed cattle to be used primarily for beef. Previously, cattle were first and foremost kept for pulling ploughs as oxen, but he crossed long-horned heifers and a Westmoreland bull to eventually create the Dishley Longhorn. As more and more farmers followed his lead, farm animals increased dramatically in size and quality. In 1700, the average weight of a bull sold for slaughter was 370 pounds (168 kg). By 1786, that weight had more than doubled to 840 pounds (381 kg). However, after his death, the Dishley Longhorn was replaced with short-horn versions.

He also bred the Improved Black Cart horse, which later became the Shire horse.

Charles Darwin coined the term 'selective breeding'; he was interested in the process as an illustration of his proposed wider process of natural selection. Darwin noted that many domesticated animals and plants had special properties that were developed by intentional animal and plant breeding from individuals that showed desirable characteristics, and discouraging the breeding of individuals with less desirable characteristics.

Darwin used the term "artificial selection" twice in the 1859 first edition of his work *On the Origin of Species*, in Chapter IV: Natural Selection, and in Chapter VI: Difficulties on Theory:

Slow though the process of selection may be, if feeble man can do much by his powers of artificial selection, I can see no limit to the amount of change, to the beauty and infinite complexity of the co-adaptations between all organic beings, one with another and with their physical conditions of life, which may be effected in the long course of time by nature's power of selection.

—Charles Darwin, On the Origin of Species

We are profoundly ignorant of the causes producing slight and unimportant variations; and we are immediately made conscious of this by reflecting on the differences in the breeds of our domesticated animals in different countries,—more especially in the less civilized countries where there has been but little artificial selection.

Animal Breeding

Three generations of "Westies" in a village in Fife, Scotland

Animals with homogeneous appearance, behavior, and other characteristics are known as particular breeds, and they are bred through culling animals with particular traits and selecting for further breeding those with other traits. Purebred animals have a single, recognizable breed, and purebreds with recorded lineage are called pedigreed. Crossbreeds are a mix of two purebreds, whereas mixed breeds are a mix of several breeds, often unknown. Animal breeding begins with breeding stock, a group of animals used for the purpose of planned breeding. When individuals are looking to breed animals, they look for certain valuable traits in purebred stock for a certain purpose, or may intend to use some type of crossbreeding to produce a new type of stock with different, and, it is presumed, superior abilities in a given area of endeavor. For example, to breed chickens, a breeder typically intends to receive eggs, meat, and new, young birds for further reproduction. Thus, the breeder has to study different breeds and types of chickens and analyze what can be expected from a certain set of characteristics before he or she starts breeding them. Therefore, when purchasing initial breeding stock, the breeder seeks a group of birds that will most closely fit the purpose intended.

Purebred breeding aims to establish and maintain stable traits, that animals will pass to the next generation. By "breeding the best to the best," employing a certain degree of inbreeding, considerable culling, and selection for "superior" qualities, one could develop a bloodline superior in certain respects to the original base stock. Such animals can be recorded with a breed registry, the organization that maintains pedigrees and/or stud books. However, single-trait breeding, breeding for only one trait over all others, can be problematic. In one case mentioned by animal behaviorist Temple Grandin, roosters bred for fast growth or heavy muscles did not know how to perform typical rooster courtship dances, which alienated the roosters from hens and led the roosters to kill the hens after reproducing with them.

The observable phenomenon of hybrid vigor stands in contrast to the notion of breed purity. However, on the other hand, indiscriminate breeding of crossbred or hybrid animals may also result in degradation of quality. Studies in evolutionary physiology, behavioral genetics, and other areas of organismal biology have also made use of deliberate selective breeding, though longer generation times and greater difficulty in breeding can make such projects challenging in vertebrates.

Advantages and Disadvantages

Selective breeding is a direct way to determine if a specific trait can "evolve in response to selection." A single-generation method of breeding is not as accurate or direct. The process is also more practical and easier to understand than sibling analysis. The former tests "differences between line means" while the latter is dependent upon "variance and covariance components." Essentially, selective breeding is better for traits such as physiology and behavior that are hard to measure because it requires fewer individuals to test than single-generation testing.

However, there are disadvantages to this process. Because a single experiment done in selective breeding cannot be used to assess an entire group of "genetic variances and covariances," individual experiments must be done for every individual trait. Also, because of the necessity of selective breeding experiments to require maintaining the organisms tested in a lab or greenhouse, it is impractical to use this breeding method on many organisms. Controlled mating instances are difficult to carry out in this case and this is a necessary component of selective breeding.

SMART BREEDING

SMART breeding (Selection with Markers and Advanced Reproductive Technologies) or Precision breeding refers to a genetic engineering technique of reproducing a species members together to retain desirable traits and so produce a stronger hybrid. The technique was successfully used by Nachum Kedar, an Israeli scientist, who applied the technique using beefsteak tomatoes to produce a fruit that would ripen on the vine and remain firm in transit.

Smart breeding has been advanced as an alternative to transgenics as a way to produce plants that are resistant to various environmental problems.

HYBRID (BIOLOGY)

In biology a hybrid, also known as cross breed, is the result of mixing, through sexual reproduction, two animals or plants of different breeds, varieties, species or genera. Using genetic terminology, it may be defined as follows:

- ***Hybrid*** generally refers to any offspring resulting from the breeding of two genetically distinct individuals, which usually will result in a high degree of heterozygosity, though hybrid and heterozygous are not, strictly speaking, synonymous;
- ***genetic hybrid*** carries two different alleles of the same gene;
- ***structural hybrid*** results from the fusion of gametes that have differing structure in at least one chromosome, as a result of structural abnormalities;
- ***numerical hybrid*** results from the fusion of gametes having different haploid numbers of chromosomes;
- ***permanent hybrid*** is a situation where only the heterozygous genotype occurs, because all homozygous combinations are lethal.

From a taxonomic perspective, hybrid refers to:

- Offspring resulting from the interbreeding between two animal species or plant species.

Hercules, a “liger”, a lion/tiger hybrid

- Hybrids between different subspecies within a species (such as between the Bengal tiger and Siberian tiger) are known as ***intra-specific*** hybrids. Hybrids between different species within the same genus (such as between lions and tigers) are sometimes known as ***interspecific*** hybrids or crosses. Hybrids between different genera (such as between sheep and goats) are known as ***intergeneric*** hybrids. Extremely rare ***interfamilial*** hybrids have been known to occur (such as the guineafowl hybrids). No ***interordinal*** (between different orders) animal hybrids are known.
- The third type of hybrid consists of crosses between populations, breeds or cultivars within a single species. This meaning is often used in plant and animal breeding, where hybrids are commonly produced and selected, because they have desirable characteristics not found or inconsistently present in the parent individuals or populations.

Terminology

The term hybrid is derived from Latin *hybrida*, meaning the ***"offspring of a tame sow and a wild boar"***, ***"child of a freeman and slave"***, etc. The term came into popular use in English in the 19th century, though examples of its use have been found from the early 17th century.

There is a popular convention of naming hybrids by forming portmanteau words. This became common in the 1920s, with the breeding of tiger-lion hybrids (liger and tigon). This was playfully (but unsystematically) extended to a number of other hybrids, or hypothetical hybrids, such as beefalo (1960s), humanzee (1980s), and cama (1998).

Types of Hybrids

Depending on the parents, there are a number of different types of hybrids;

- ***Single cross hybrids*** result from the cross between two true breeding organisms which produces an F1 generation called an F1 hybrid (F1 is short for Filial 1, meaning «first offspring»). The cross between two different homozygous lines produces an F1 hybrid that is heterozygous; having two alleles, one contributed by each parent and typically one is dominant and the other recessive. Typically, the F1 generation is also phenotypically homogeneous, producing offspring that are all similar to each other.
- ***Double cross hybrids*** result from the cross between two different F1 hybrids.
- ***Three-way cross hybrids*** result from the cross between an F1 hybrid and an inbred line.
- ***Triple cross hybrids*** result from the crossing of two different three-way cross hybrids.

- ***Population hybrids*** result from the crossing of plants or animals in a population with those of another population. These include crosses between organisms such as interspecific hybrids or crosses between different breeds.
- ***Stable hybrid*** is a horticultural term, usually referring to an annual plant that, if grown and bred in a small monoculture free of external pollen (e.g., an air-filtered greenhouse) will produce offspring that are "true to type" with respect to phenotype; i.e., a true breeding organism.
- ***Hybrid species*** result from hybrid populations evolving reproductive barriers against their parent species through hybrid speciation.

Interspecific Hybrids

Interspecific hybrids are bred by mating two species, normally from within the same genus. The offspring display traits and characteristics of both parents. The offspring of an interspecific cross are often sterile; thus, hybrid sterility prevents the movement of genes from one species to the other, keeping both species distinct. Sterility is often attributed to the different number of chromosomes between the two species. For example, donkeys have 62 chromosomes, while horses have 64 chromosomes, and mules and hinnies have 63 chromosomes. Mules, hinnies, and other normally sterile interspecific hybrids cannot produce viable gametes, because differences in chromosome structure prevent appropriate pairing and segregation during meiosis, meiosis is disrupted, and viable sperm and eggs are not formed. However, fertility in female mules has been reported with a donkey as the father.

Most often other processes occurring in plants and animals maintain gametic isolation and species distinction. Species often have different mating or courtship patterns or behaviors, the breeding seasons may be distinct and even if mating does occur antigenic reactions to the sperm of other species prevent fertilization or embryo development. Hybridisation is much more common among organisms that spawn indiscriminately, like soft corals and among plants.

While it is possible to predict the genetic composition of a backcross *on average*, it is not possible to accurately predict the composition of a particular backcrossed individual, due to random segregation of chromosomes. In a species with two pairs of chromosomes, a twice backcrossed individual would be predicted to contain 12.5% of one species' genome (say, species A). However, it may, in fact, still be a 50% hybrid if the chromosomes from species A were lucky in two successive segregations, and meiotic crossovers happened near the telomeres. The chance of this is fairly high: $\left(\frac{1}{2}\right)^{(2\times2)} = \frac{1}{16}$ (where the "two times two" comes about from two rounds of meiosis with two chromosomes); however, this probability declines markedly with chromosome number and so the actual composition of a hybrid will be increasingly closer to the predicted composition.

HYBRID SPECIATION

Hybrid speciation is a form of speciation wherein hybridization between two different species leads to a new species, reproductively isolated from the parent species. From the 1940s, reproductive isolation between hybrids and their parents was thought to be particularly difficult to achieve and thus hybrid species were thought to be extremely rare. With DNA analysis becoming more accessible in the 1990s, hybrid speciation has been shown to be a fairly common phenomenon, particularly in plants.

In botanical nomenclature, a hybrid species is also called a nothospecies. Hybrid species are by their nature polyphyletic.

Hybrid Speciation Ecology

A hybrid may have a distinct trait (phenotype). This phenotype may in very rare cases be better fitted to the local environment than the parental lineage and as such natural selection may favor these individuals. If reproductive isolation subsequently is achieved, it will lead to a separate species. The reproductive isolation may be genetic, ecological, behavioural, or spatial, or a combination of these.

If reproductive isolation fails to establish, the hybrid population may breed back and finally merge with either or both parent species. This will lead to an influx of foreign genes in the parent population, a situation called an introgression. Introgression is a source of genetic variation, and can in itself facilitate speciation. There is evidence that introgression is a ubiquitous phenomenon in plants, animals, and even humans, where genetic material from Neanderthals and Denisovans is responsible for much of the immune genes in non-African populations.

Ecological Constraints to Hybrid Speciation

For a hybrid form to persist, it will generally have to be able to exploit the available resources better than either parent species, which, in most cases, it will have to compete with. While grizzly bears and polar bears may have offspring, a grizzly–polar bear hybrid will likely be less suited in either of the ecological roles than the parents themselves. Although the hybrid is fertile, this poor adaptation would prevent the establishment of a permanent population.

Likewise, lions and tigers have historically overlapped in a portion of their range and can theoretically produce wild hybrids: ligers, which are a cross between a male lion and female tiger, and tigons, which are a cross between a male tiger and a female lion; however, tigers and lions have thus far only hybridized in captivity. In both ligers and tigons, the females are fertile and the males are sterile. One of these hybrids (the tigon) carries growth-inhibitor genes from both parents and thus is smaller than either parent species and might in the wild come into competition with smaller carnivores, e.g. the leopard. The other hybrid, the liger, ends up larger than either of its parents: about a thousand pounds (450 kg) fully-grown. No tiger-lion hybrids are known from the wild, particularly because each species is confined to geographically separated ranges (tigers aren't found in Africa, and the Asiatic lion is only found in the Gir Forest National Park, where tigers also are absent).

Some situations may favour hybrid population. One example is rapid turnover of available environment types, like the historical fluctuation of water level in Lake Malawi, a situation that generally favors speciation. A similar situation can be found where closely related species occupy a chain of islands. This will allow any present hybrid population to move into a new unoccupied habitats, avoiding direct competition with parent species and giving a hybrid population time and space to establish. Genetics too can occasionally favour hybrids. In the Amboseli National Park in Kenya, yellow baboons and anubis baboons regularly interbreed. The hybrid males reach maturity earlier than their pure bred cousins, setting up a situation where the hybrid population may over time replace one or both of the parent species in the area.

Genetics of Hybridization

Genetics are more variable and malleable in plants than in animals, probably reflecting the higher activity level in animals. Hybrids genetics will necessarily be less stable than those of species evolving through isolation, which explains why hybrid species appear more common in plants than in animals. Many agricultural crops are hybrids with double or even triple chromosome sets. Having multiple sets of chromosomes is called polyploidy (or polyploidity). Polyploidy is usually fatal in animals where extra chromosome sets upset fetal development, but is often found in plants. A form of hybrid speciation that is relatively common in plants, occurs when an infertile hybrid becomes fertile after doubling of the chromosome number.

Hybridization without change in chromosome number is called homoploid hybrid speciation. This is the situation found in most animal hybrids. For a hybrid to be viable, the chromosomes of the two organisms will have to be very similar, i.e., the parent species must be closely related, or the difference in chromosome arrangement will make mitosis problematic. With polyploid hybridization, this constraint is less acute.

Super-numerary chromosome numbers can be unstable, which can lead to instability in the genetics of the hybrid. The European edible frog appears to be a species, but is actually triploid semi-permanent hybrids between pool frogs and marsh frogs. In most populations, the edible frog population is dependent on the presence of at least one of the parents species to be maintained as each individual need two gene sets from one parent species and one from the other. Also, the male sex determination gene in the hybrids is only found in the genome of the pool frog, further undermining stability. Such instability can also lead to rapid reduction of chromosome numbers, creating reproductive barriers and thus allowing speciation.

Known cases of Hybrid Speciation

Closely related *Heliconius* species

Animals

Hybrid speciation in animals is primarily homoploid. While not very common, a few animal species have been recognized as being the result of hybridization, mostly insects and fish. The Lonicera fly is an example of a novel animal species that resulted from natural hybridization. The great skua has a surprising genetic similarity to the physically dissimilar pomarine skua, and most ornithologists now assume the great skua is a hybrid species between the pomarine skua and one of the northern skua species.

Rapidly diverging species can sometimes form multiple hybrid species, giving rise to a species complex, like several physically divergent but closely related genera of cichlid fishes in Lake Malawi. The duck genus *Anas* (mallards and teals) has a very recent divergence history, many of the species are inter-fertile and quite a few of them are thought to be hybrids. While hybrid species generally appear rare in mammals, the American red wolf too appear to be a hybrid species of the *Canis* species complex, between gray wolf and coyote. This is also hypothesized to have led to the species rich *Heliconius* butterflies, though the conclusion has been criticized.

The first known instance of hybrid speciation in marine mammals was discovered by scientists in 2014. The clymene dolphin (*Stenella clymene*), also known in older texts as the short-snouted spinner dolphin, is a hybrid of the spinner dolphin and striped dolphin. All three species live in the Atlantic Ocean and often interact peacefully.

Examples of Hybrid Animals and Animal Populations Derived from Hybrids

Mammals

A "zonkey", a zebra/donkey hybrid

A "jaglion", a jaguar/lion hybrid

- Equid hybrids
- Mule, a cross of female horse and a male donkey

- Hinny, a cross between a female donkey and a male horse (Mule and hinny are examples of reciprocal hybrids)
- Zebroids
- Zeedonk or Zonkey, a zebra/donkey cross
- Zorse, a zebra/horse cross
- Zony or Zetland, a zebra/pony cross ("zony" is a generic term; "zetland" is specifically a hybrid of the Shetland pony breed with a zebra)
- hybrid ass, a cross between a donkey and an onager or Asian wild ass
- Bovid hybrids
- Dzo, zo or yakow; a cross between a domestic cow/bull and a yak
- Beefalo, a cross of an American bison and a domestic cow. This is a fertile breed; this along with genetic evidence has caused them to be recently reclassified into the same genus, *Bos*.
- Żubroń, a hybrid between wisent (European bison) and domestic cow
- «Higgs bison», a naturally occurring hybrid between aurochs and steppe bison that appeared during the Pleistocene; likely gave rise to the wisent.
- Sheep-goat hybrid is the cross between a sheep and a goat, which belong to different genera.
- Ursid hybrids, such as the grizzly-polar bear hybrid, occur between black bears, brown bears, and polar bears.
- Felid hybrids
- Savannah cat is a fertile ***breed*** developed originally from a cross between the serval (*Leptailurus serval*) and domestic cats (*Felis catus*).
- A hybrid between a Bengal tiger and a Siberian tiger is an example of an ***intra-specific*** hybrid. It also includes the Indochinese tiger, Sumatran tiger.
- Pumapards are the hybrid crosses between a puma and a leopard.
- Ligers and tigons (crosses between a lion and a tiger - the difference in name due to which parent is of which species - ligers have a lion father and a tiger mother) and other Panthera hybrids such as the lijagulep. Various other wild cat crosses are known involving the lynx, bobcat, leopard, serval, etc.
- Liligers are the hybrid cross between a male lion and a ligress.
- Bengals are a fertile ***breed*** developed originally from a cross between the Asian leopard cat *Prionailurus bengalensis* and the domestic cat *Felis catus*.
- Fertile canid hybrids occur between coyotes, wolves, dingoes, jackals and domestic dogs.
- Hybrids between black and white rhinoceroses have been recognized.
- Hybrid camel, a cross between a bactrian camel and a dromedary camel
- Cama, a cross between a camel and a llama, also an intergeneric hybrid.
- Wholphin, a fertile but very rare cross between a false killer whale and a

bottlenose dolphin.

- At Chester Zoo in the United Kingdom, a cross between an African elephant (male) and an Asian elephant (female). The male calf was named Motty. He died of intestinal infection after twelve days.
- Bornean and Sumatran orangutan hybrids have occurred in captivity.

Birds

A mule, a domestic canary/goldfinch hybrid

- Hybrids between spotted owls and barred owls
- Cagebird breeders sometimes breed hybrids between species of finch, such as goldfinch × canary. These birds are known as mules.
- The perlin is a peregrine falcon – merlin hybrid.
- Gamebird hybrids, hybrids between gamebirds and domestic fowl, including chickens, guineafowl and peafowl, interfamilial hybrids.
- Numerous macaw hybrids and lovebird hybrids are also known in aviculture.
- Red kite × black kite: five bred unintentionally at a falconry center in England. (It is reported that the black kite (the male) refused female black kites but mated with two female red kites.)
- The mulard duck, hybrid of the domestic pekin duck and domesticated muscovy ducks.
- In Australia, New Zealand and other areas where the Pacific black duck occurs, it is hybridised by the much more aggressive introduced mallard. This is a concern to wildlife authorities throughout the affected area, as it is seen as Genetic pollution of the black duck gene pool.
- Hybridisation in gulls is a reasonably frequent occurrence in the wild.

Reptiles

- Hybrid iguana, a single-cross hybrid resulting from natural interbreeding between male marine iguanas and female land iguanas since the late 2000s.

- Crestoua, a cross between a Rhacodactylus Ciliatus (crested gecko) and a Rhacodactylus Chahoua.
- Colubrid snakes of the tribe Lampropeltini have been shown to produce fertile hybrid offspring.
- Hybridization between the endemic Cuban crocodile (*Crocodilus rhombifer*) and the widely distributed American crocodile (*Crocodilus acutus*) is causing conservation problems for the former species as a threat to its genetic integrity.
- Saltwater crocodiles (*Crocodylus porosus*) have mated with Siamese crocodiles (*Crocodylus siamensis*) in captivity producing offspring which in many cases have grown over 20 feet (6.1 metres) in length. It is likely that wild hybridization occurred historically in parts of southeast Asia.
- Many species of boas and pythons are known to produce hybrids, such as carball (a cross between a ball python and a carpet python) or a bloodball (a cross between a blood python and a ball python); however most of these only occur in captivity. Contrary to popular belief, boa–python hybrids are not possible due to their differing reproductive functions. Boas only produce hybrids with other species of boas, and pythons only produce hybrids with other species of pythons.

Amphibians

- Japanese giant salamanders and Chinese giant salamanders have created hybrids that threaten the survival of Japanese giant salamanders due to the competition for similar resources in Japan.

Fish

- Blood parrot cichlid, which is probably created by crossing a red head cichlid and a Midas cichlid or red devil cichlid.
- A group of about 50 hybrids between Australian blacktip shark and the larger common blacktip shark was found by Australia's East Coast in 2012. This is the only known case of hybridization in sharks.
- Silver bream and common bream commonly produce sterile hybrids.
- Tiger muskie is a sterile hybrid between northern pike and muskellunge.

Insects

- Killer bees were created during an attempt to breed a strain of bees that would produce more honey and be better adapted to tropical conditions. This was done by crossing a European honey bee and an African bee.
- The *Colias eurytheme* and *C. philodice* butterflies have retained enough genetic compatibility to produce viable hybrid offspring.

Species	Common name	Family	Hybridization	Confirmed or putative hybridization?	Putative parental/ introgressive species	Polyploid or homoploid?	Polyploid chromosome count	References	Notes
Abelmoschus esculentus (L.) Moench	Okra	Malvaceae	Allopolyploid origin	Putative	Uncertain	Polyploid (tetraploid)	usually 2n=4x=130	Joshi and Hardas, 1956; Schafleitner et al., 2013	Variable ploidy
Actinidia deliciosa (A. Chev.) C.F.Liang & A.R.Ferguson	Kiwifruit	Actinidiaceae	Allopolyploid origin	Putative	Actinidia chinensis Planch. and Unknown	Polyploid (hexaploid)	2n=6x=174	Atkinson et al., 1997	
Agave fourcroydes Lem.	Henequen	Asparagaceae	Allopolyploid origin	Confirmed	Uncertain	Polyploid (usu. pentaploid, triploid)	2n=5x(3x)=150(90)	Robert et al., 2008; Hughes et al., 2007	Variable ploidy, polyploid event not recent
Agave sisalana Perrine	Sisal	Asparagaceae	Allopolyploid origin	Confirmed	Uncertain	Polyploid (usu. pentaploid, hexaploid)	2n=5x(6x)=150(180)	Robert et al., 2008	Variable ploidy, polyploid event not recent
Allium ampeloprasum L.	Great headed garlic	Amaryllidaceae	Intraspecific hybrid origin	Putative	Allium ampeloprasum L.	Homoploid	–	Guenaoui et al., 2013	
Allium cepa L.	Common onion	Amaryllidaceae	Interspecific hybrid origin	Putative	Uncertain: Allium vavilovii Popov & Vved., A. galanthum Kar. & Kir. or A. fistulosum L.	Homoploid	–	Gurushidze et al., 2007	
Allium cornutum Clementi	Triploid onion	Amaryllidaceae	Triparental allopolyploid origin	Confirmed	Allium cepa L., A. roylei Stearn, unknown	Polyploid (triploid)	2n=3x=24	Fredotovic et al., 2014	
Ananas comosus (L.) Merr is	Pineapple	Bromeliaceae	Interspecific introgression	Putative	Ananas ananassoides (Baker) L.B. Smith	Homoploid	–	Duval et al., 2003	
Annona x atemoya	Atemoya	Annonaceae	Interspecific hybrid origin	Confirmed	Annona cherimola Mill. and A. squamosa L.	?	–	Perfectti et al., 2004; Jalikop, 2010	

Arachis hypogaea L.	Peanut	Fabaceae	Allopolyploid origin	Confirmed	Arachis duranensis Krapov. & W.C. Greg. and A. ipaënsis Krapov. & W.C. Greg.	Polypoid (tetraploid)	2n=4x=40	Kochert et al., 1996; Bertioli et al., 2011	
Armoracia rusticana P.Gaertn., B.Mey. & Scherb.	Horseradish	Brassicaceae	Interspecific hybrid origin	Putative	Uncertain	?	–	Courter and Rhodes, 1969	
Artocarpus altilis (Parkinson ex F.A.Zorn) Fosberg	Breadfruit	Moraceae	Interspecific introgression	Putative	Artocarpus mariannensis Trécul	?	–	Zerega et al., 2005; Jones et al., 2013	
Avena sativa L.	Oat	Poaceae	Allopolyploid origin	Confirmed	Uncertain	Polyploid (hexaploid)	2n=6x=42	Linares et al., 1998; Oliver et al., 2013	
Brassica carinata A.Braun	Ethiopian mustard	Brassicaceae	Allopolyploid origin	Confirmed	Brassica oleracea L. and B. nigra (L.) K.Koch	Polyploid (tetraploid)	2n=4x=19	Arias et al., 2014	
Brassica juncea (L.) Czern.	Indian mustard	Brassicaceae	Allopolyploid origin	Confirmed	Brassica nigra (L.) K.Koch and B. rapa L.	Polyploid (tetraploid)	2n=4x=18	Arias et al., 2014	
Brassica napus L.	Rapeseed, rutabega	Brassicaceae	Allopolyploid origin	Confirmed	Brassica rapa L. and B. oleracea L.	Polyploid (tetraploid)	2n=4x=19	Arias et al., 2014	
Cajanus cajan (L.) Millsp.	Pigeon pea	Fabaceae	Intraspecific introgression, interspecific introgression	Putative	Wild Cajanus cajan and other species	Homoploid	–	Kassa et al., 2012	
Cannabis sativa L.	Hemp	Cannabaceae	Intraspecific introgression	Putative	Cannabis sativa L. ‘Indica’ and ‘Sativa’ types	Homoploid	–	de Meijer and van Soest, 1992	
Carica pentagona Heilborn	Babaco	Caricaceae	Interspecific hybrid origin	Confirmed	Uncertain (Carica stipulata V.M.Badillo, Vasconcellea pubescens A.DC., Vasconcellea weberbaueri (Harms) V.M. Badillo)	?	–	Van Droogenbroeck et al., 2002; Van Droogenbroeck et al., 2006	

Carya illinoinensis (Wangenh.) K.Koch	Pecan	Juglandaceae	Interspecific hybrid origin	Putative	Uncertain	?	–	Grauke et al., 2011	
Castanea dentata (Marshall) Borkh	Chestnut	Fagaceae	Interspecific introgression	Confirmed	Castanea pumila (L.) Mill.	Homoploid	–	Li and Dane, 2013	Also ongoing efforts to introgress blight resistance from Castanea mollissima Blume (see Jacobs et al., 2013)
Castanea sativa Mill.	Chestnut	Fagaceae	Intraspecific introgression	Confirmed	Castanea sativa Eurosiberian and Mediterranean populations	Homoploid	–	Villani et al., 1999; Mattioni et al., 2013	
Chenopodium quinoa Willd.	Quinoa	Chenopodiaceae	Allopolyploid origin	Putative	Uncertain	Polyploid (tetraploid)	–	Heiser, 1974; Ward, 2000; Maughan et al., 2004	
Cicer arietinum L.	Chickpea (pea-shaped)	Fabaceae	Intraspecific hybrid origin	Putative	Cicer arietinum L. Desi and Kabuli Germplasm	?	–	Upadhyaya et al., 2008; Keneni et al., 2011	
Cichorium intybus L.	Radicchio	Asteraceae	Interspecific introgression	Confirmed	Wild Cichorium intybus L.	Homoploid	–	Kiaer et al., 2009	
Citrus aurantiifolia (Christm.) Swingle	Key lime	Rutaceae	Interspecific hybrid origin	Confirmed	Citrus medica L. and C. subg. Papeda	?	–	Ollitrault and Navarro, 2012; Penjor et al., 2014; Nicolosi et al., 2000; Moore, 2001	
Citrus aurantium L.	Sour oranges	Rutaceae	Interspecific hybrid origin	Confirmed	Citrus maxima (Burm.) and C. reticulata Blanco	?	–	Wu et al., 2014; Moore, 2001	
Citrus clementina hort.	Clementine	Rutaceae	Interspecific hybrid origin	Confirmed	Citrus sinensis (L.) Osbeck and C. reticulata Blanco	?	–	Wu et al., 2014	
Citrus limon (L.) Osbeck	Lemon, lime	Rutaceae	Interspecific hybrid origin	Confirmed	Citrus medica L., C. aurantiifolia (Christm.) Swingle, and uncertain	?	–	Nicolosi et al., 2000; Moore, 2001	
Citrus paradisi Macfad.	Grapefruit	Rutaceae	Interspecific hybrid origin	Confirmed	Citrus sinensis (L.) Osbeck and C. maxima (Burm.)	?	–	Wu et al., 2014; Moore, 2001	

Citrus reticulata Blanco	Mandarin	Rutaceae	Interspecific introgression	Confirmed	Citrus maxima (Burm.)	?	–	Wu et al., 2014	
Citrus sinensis (L.) Osbeck	Sweet orange (blood, common)	Rutaceae	Interspecific hybrid origin	Confirmed	Uncertain	?	–	Wu et al., 2014; Moore, 2001	
Cocos nucifera L.	Coconut	Arecaceae	Intraspecific introgression	Confirmed	Cocos nucifera L. Indo-Atlantic and Pacific lineages	Homoploid	–	Gunn et al., 2011	
Coffea arabica L.	Coffee	Rubiaceae	Allopolyploid origin	Confirmed	Coffea eugenioides S.Moore and C. canephora Pierre ex A.Froehner	Polyploid (tetraploid)	2n=4x=44	Lashermes et al., 1999	
Corylus avellana L.	Hazelnut	Betulaceae	Intraspecific introgression	Confirmed	Wild Corylus avellana L. in Southern Europe	Homoploid	–	Campa et al., 2011; Boccacci et al., 2013	
Cucurbita pepo L.	Winter squash, pumpkin	Cucurbitaceae	Intraspecific introgression	Putative	Cucurbita pepo var. texana (Scheele) D.S.Decker	Hompolid	–	Kirkpatrick and Wilson, 1988	
Daucus carota subsp. sativus (Hoffm.) Arcang.	Carrot	Apiaceae	Intraspecific introgression	Confirmed	Daucus carota L. subsp. carota	Homoploid	–	Iorizzo et al., 2013; Simon, 2000	
Dioscorea L. spp.	Yam	Dioscoreaceae	Interspecific hybrid origin, introgression	Putative	Uncertain	Variable	–	Terauchi et al., 1992; Dansi et al., 1999; Bhattacharjee et al., 2011; Mignouna et al., 2002	Multiple species of putative hybrid (perhaps allopolyploid) origin including Dioscorea cayennensis subsp. rotundata (Poir.) J.Miège. and D. cayennensis Lam.
Diospyros kaki L.f.	Persimmon	Ebenaceae	Allopolyploid origin	Putative	Uncertain	Polyploid (hexaploid)	Yonemori et al., 2008		
Ficus carica L.	Fig	Moraceae	Interspecific introgression	Putative	Uncertain	?	Aradhya et al., 2010		

Fragaria ananassa (Duchesne ex Weston) Duchesne ex Rozier	Strawberries	Rosaceae	Interspecific hybrid origin	Confirmed	Fragaria virginiana Mill. (octoploid), F. chiloensis (L.) Mill. (octoploid)	Homoploid relative to parentals (octoploid)	2n=8x=56 Evans, 1977; Hirakawa et al., 2014	Uncertain which species formed the octoploid progenitors	
Garcinia mangostana L.	Mangosteen	Clusiaceae	Allopolyploid origin	Putative	Garcinia celebica L. and G. malaccensis Hook.f. ex T.Anderson	Polyploid (tetraploid)	? Richards, 1990	Recent work shows this may not be of hybrid origin (Nazre, 2014)	
Gossypium hirsutum L.	Upland cotton	Malvaceae	hybrid origin	Confirmed	Uncertain, referred to as 'A' and 'D'	Polyploid (formed <1MYA)	2n =4x=52 Wendel and Cronn 2003	Polyploidization likely led to agronomically significant traits (Applequist et al., 2001)	
Hibiscus sabdariffa L.	Roselle	Malvaceae	Allopolyploid origin	Putative	Uncertain	Polyploid (tetraploid)	2n=4x=72 Menzel and Wilson, 1966; Satya et al., 2013		
Hordeum vulgare L.	Barley	Poaceae	Introgression	Confirmed	Hordeum spontaneum K.Koch	Homoploid	Badr et al., 2000; Dai et al., 2012		
Humulus lupulus L.	Hops	Cannabaceae	Intraspecific introgression	Confirmed	Humulus lupulus L. North American and European Germplasm	Homoploid	Reeves and Richards, 2011; Stajner et al., 2008; Seefelder et al., 2000		
Ipomoea batatas (L.) Lam.	Sweet potato	Convolvulaceae	Intraspecific introgression; Interspecific introgression?	Putative	Ipomoea batatas (L.) Lam. Central American and South American Germplasm	Homoploid relative to parentals	Roullier et al., 2013		
Juglans regia L.	Walnut	Juglandaceae	Interspecific hybridization	Confirmed	Juglans sigillata Dode	Homoploid	Gunn et al., 2010		
Lactuca sativa L.	Lettuce	Asteraceae	Intraspecific hybrid origin	Putative	Lactuca serriola L. and other L. spp.	Homoploid	de Vries, 1997		
Lagenaria siceraria (Molina) Standl.	Bottle Gourd	Cucurbitaceae	Intraspecific introgression	Confirmed	Lagenaria siceraria (Molina) Standl. African/American and Asian Germplasm	Homoploid	Clarke et al., 2006		
Lens culinaris Medik. ssp. culinaris	Lentil	Fabaceae	Intraspecific introgression	Putative	Wild lentil, Lens culinaris subsp. orientalis (Boiss.) Ponert	Homoploid	Erskine et al., 2011		

Macadamia integrifolia Maiden & Betche	Macadamia	Proteaceae	Interspecific hybrid origin, interspecific introgression	Confirmed	Macadamia tetraphylla L.A.S.Johnson, and other M. spp.	Homoploid	Hardner et al., 2009; Steiger et al., 2003; Aradhya et al., 1998		
Malus domestica Borkh.	Apple	Rosaceae	Interspecific hybrid origin	Confirmed	Malus sieversii (Ledeb.) M.Roem., M. sylvestris (L.) Mill., and possibly others	Homoploid	Cornille et al., 2012		
Mentha piperita L.	Peppermint	Lamiaceae	Allopolyploid origin	Confirmed	Mentha aquatica L. and M. spicata L.	Polyploid (12-ploid)	2n=12x=66 or 72 Harley and Brighton, 1977; Gobert et al., 2002		
Musa paradisiaca L.	Banana	Musaceae	Allopolyploid origin	Confirmed	Musa acuminata Colla, M. balbisiana Colla	Polyploid (usually triploid)	2n=3x=33 Simmonds and Shepherd, 1955; Heslop-Harrison and Schwarzacher, 2007; De Langhe et al., 2010		
Nicotiana tabacum L.	Tobacco	Solanaceae	Allopolyploid origin	Confirmed	Uncertain (Nicotiana sylvestris Speg. & S. Comes and N. tomentosiformis Goodsp.)	Polyploid (tetraploid)	2n=4x=48 Kenton et al., 1993; Murad et al., 2002		
Olea europaea L.	Olive	Oleaceae	Intraspecific introgression	Putative	Wild Olea europaea L., Eastern and Western Germplasm	Homoploid	Kaniewski et al., 2012; Besnard et al., 2013; Breton et al., 2006; Rubio de Casas et al., 2006; Besnard et al., 2007; Besnard et al., 2000		
Opuntia L. spp.	Opuntia	Cactaceae	Interspecific hybrid origin, Allopolyploid origin	Putative	Including Opuntia ficus-indica (L.) Mill.	Polyploid, homoploid	Hughes et al., 2007; Griffith, 2004		
Oryza sativa L.	Rice	Poaceae	Intraspecific introgression, interspecific introgression	Putative	Oryza sativa L. 'Japonica' and 'Indica' Germplasm, Oryza rufipogon Griff.	Homoploid	Caicedo et al., 2007; Gao and Innan, 2008		

Oxalis tuberosa Molina	Oca	Oxalidaceae	Allopolyploid origin	Putative	Uncertain	Polyploid (octaploid)	2n=8x=64 Emswhiller and Doyle, 2002; Emshwiller 2002; Emswhiller et al., 2009		
Pennisetum glaucum (L.) R.Br.	Pearl millet	Poaceae	Intraspecific introgression	Confirmed	Wild Pennisetum glaucum (L.) R.Br.	Homoploid	Oumar et al., 2008		
Persea americana Mill.	Avocado (Hass and other cultivars)	Lauraceae	Intraspecific introgression	Confirmed	Persea americana Mill. 'Guatamalensis', 'Drymifolia', and' Americana'	Homoploid	Chen et al., 2008; Davis et al., 1998; Ashworth and Clegg, 2003; Douhan et al., 2011		
Phoenix dactylifera L.	Date palm	Arecaceae	Interspecific hybrid origin	Putative	Uncertain	Homoploid	El Hadrami et al., 2011; Bennaceur et al., 1991		
Piper methysticum G.Forst.	Kava	Piperaceae	Allopolyploid origin	Putative	Piper wichmannii C. DC. and P. gibbiflorum C.DC.	Polyploid (decaploid)	2n=10x=130 Singh, 2004; Lebot et al., 1991		
Pistacia vera L.	Pistachio	Anacardiaceae	Interspecific introgression	Putative	Pistacia atlantica Desf., P. chinensis subsp. integerrima (J. L. Stewart ex Brandis) Rech. f.	Homoploid	Kafkas et al., 2001		
Pisum abyssinicum A.Braun	Pea	Fabaceae	Interspecific hybrid origin	Confirmed	Uncertain (Pisum fulvum Sibth. & Sm. and other P. spp.)	Homoploid	Vershinin et al., 2003		
Pisum sativum L.	Pea	Fabaceae	Interspecific hybrid origin	Confirmed	Uncertain (Pisum sativum subsp. elatius (M.Bieb.) Asch. & Graebn. and other P. spp.)	Homoploid	Vershinin et al., 2003		
Plinia cauliflora (Mart.) Kausel	Jaboticaba	Myrtaceae	Intraspecific hybridization, interspecific hybridization	Putative	Plinia 'Jaboticaba' and 'Cauliflora' Germplasm; P. peruviana (Poir.) Govaerts	Homoploid	Balerdi et al., 2006		

Prunus cerasus L.	Cherry	Rosaceae	Allopolyploid origin	Confirmed	Prunus avium (L.) L. and P. fruticosa Pall.	Polyploid (tetraploid)	2n=4x=32 Tavaud et al., 2004; Olden and Nybom, 1968		
Prunus domestica L.	Plum	Rosaceae	Allopolyploid origin	Confirmed	Uncertain (P. cerasifera Ehrh. and P. spinosa L.)	Polyploid (hexaploid)	2n=6x=48 Zohary, 1992; Hartmann and Neumuller, 2009	Japanese plum is also of hybrid origin (see Hartmann and Neumuller 2009). Also hybridizes with other cultivated Prunus spp.	
Prunus dulcis (Mill.) D.A.Webb	Almond	Rosaceae	Interspecific introgression	Confirmed	Prunus orientalis (Mill.) Koehne and other P. spp.	Homoploid	Delplancke et al., 2012; Delplancke et al., 2013		
Pyrus L. species	Pear	Rosaceae	Interspecific hybrid origin	Confirmed	Many species	Homoploid	Silva et al., 2014	Also introgression with semidomesticated populations (see Iketani et al. 2009)	
Raphanus raphanistrum subsp. sativus (L.) Domin	Radish	Brassicaceae	Intraspecific introgression	Confirmed	Raphanus raphanistrum L. subsp. raphanistrum	Homoploid	Ridley et al., 2008		
Rheum L. cultivated species	Rhubarb	Polygonaceae	Interspecific hybrid origin	Putative	Unclear	Homoploid relative to parentals (tetraploid)	Foust and Marshall, 1991; Kuhl and Deboer, 2008	Hybrids include: Rheum rhaponticum L., R. rhabarbarum L., R. palmatum L.	
Rubus L. spp.	Red raspberry, blackberry, tayberry, boysenberry, etc.	Rosaceae	Allopolyploid origin, interspecific hybridization	Confirmed	Many	Polyploid	Alice et al., 2014; Alice and Campbell, 1999		
Saccharum spp.	Sugarcane	Poaceae	Allopolyploid origin	Confirmed	Saccharum officinarum L. and S. spontaneum L.	Polyploid	Variable, 2n=10-13x=100-130 Grivet et al., 1995; D'Hont et al., 1996		
Secale cereale L.	Rye	Poaceae	Interspecific hybrid origin	Confirmed	Uncertain (Secale montanum Guss., S. vavilovii Grossh.)	Homoploid	Bartos et al., 2008; Korzun et al., 2001; Hillman, 1978; Tang et al., 2011; Salamini et al., 2002		

Sechium edule (Jacq.) Sw.	Chayote	Cucurbitaceae	Interspecific introgression	Putative	Sechium compositum (Donn. Sm.) C. Jeffrey	Homoploid	Newstrom, 1991		
Setaria italica (L.) P.Beauv.	Foxtail millet	Poaceae	Interspecific introgression	Confirmed	Setaria viridis (L.) P.Beauv.	Homoploid	Till-Bottraud et al., 1992		
Solanum L. spp. Section Petota	Potato	Solanaceae	Interspecific hybrid origin, allopolyploid origin, interspecific introgression	Confirmed	Including Solanum tuberosum L., S. ajanhuiri Juz. & Bukasov, S. curtilobum Juz. & Bukasov, S. juzepczukii Bukasov	Homoploid and Polyploid	Rodriguez et al., 2010		
Solanum lycopersicum L.	Tomato	Solanaceae	Intraspecific introgression, interspecific introgression	Confirmed	Solanum lycopersicum var. cerasiforme (Dunal) D.M. Spooner, G.J. Anderson & R.K. Jansen and S. pimpinellifolium L.	Homoploid	Blanca et al., 2012; Causse et al., 2013; Rick, 1950		
Solanum melongena L.	Eggplant	Solanaceae	Interspecific hybrid origin, interspecific introgression, intraspecific introgression	Confirmed	Solanum undatum Lam. and others; wild S. melongena L. (=S. insanum L.)	Homoploid	Knapp et al., 2013; Meyer et al., 2012	Hybrid origin is not confirmed, but introgression is well documented	
Solanum muricatum Aiton	Pepino dulce	Solanaceae	Interspecific hybrid origin, interspecific introgression	Likely	Solanum species in Series Caripensia	Homoploid	Blanca et al., 2007	Polyphyletic origin and extensive, ongoing introgression with wild species	
Solanum tuberosum	Potato	Solanaceae	Intraspecific Hybrid	Confirmed	Diploid potato lines crossed with *S. chacoense* containing Sli gene	Homoploid	Lindhout et al., 2011		
Spondias purpurea L.	Jocote	Anacardiaceae	Interspecific introgression	Putative	Spondias mombin L.	Homoploid	Miller and Schaal, 2005		

Theobroma cacao L.	Cacao (Trinitario-type)	Malvaceae	Intraspecific hybrid origin	Confirmed	Theobroma cacao L. 'Forastero' and 'Criollo' Germplasm	Homoploid	Yang et al., 2013		
Triticum aestivum L.	Bread wheat, spelt	Poaceae	Allopolyploid origin	Confirmed	Triticum turgidum L. (tetraploid) with Aegilops tauschii Coss.	Polyploid (hexaploid)	2n=6x=42 Matsuoka, 2011; Dvorak, 2012		
Triticum turgidum L.	Emmer wheat, durum wheat	Poaceae	Allopolyploid origin	Confirmed	Triticum urartu Thumanjan ex Gandilyan and Aegilops speltoides Tausch	Polyploid (tetraploid)	2n=4x=28 Dvorak et al., 2012; Matsuoka, 2011; Yamane and Kawahara, 2005		
Vaccinium corymbosum L.	Highbush blueberry	Ericaceae	Interspecific hybrid origin, interspecific introgression	Putative	Vaccinium tenellum Aiton, V. darrowii Camp, (V. virgatum Aiton, V. angustifolium Aiton)	Uncertain	Vander Kloet, 1980; Bruederle et al., 1994; Lyrene et al., 2003; Boches et al., 2006	Possible hybrid origin during the Plesitocene	
Vanilla tahitensis J.W. Moore	Tahitian vanilla	Orchidaceae	Allopolyploid origin	Confirmed	Vanilla planifolia Jacks. ex Andrews and V. odorata C.Presl	Polyploid	Variable, 2n=2x(4x)=32(64) Lubinsky et al., 2008		
Vitis rotundifolia Michx.	Grape	Vitaceae	Interspecific introgression	Confirmed	Many Vitis spp.	Homoploid	2n=2x=38 Reisch et al., 2012; This et al., 2006		
Zea mays L.	Maize	Poaeceae	Intraspecific introgression	Confirmed	Wild Zea mays L. (teosinte, =subsp. parviglumis Iltis & Doebley)	Homoploid	Van Heerwaarden et al., 2011; Hufford et al., 2013		

Sterility in a non-polyploid hybrid is often a result of chromosome number; if parents are of differing chromosome pair number, the offspring will have an odd number of chromosomes, leaving them unable to produce chromosomally balanced gametes. While this is undesirable in a crop such as wheat, where growing a crop which produces no seeds would be pointless, it is an attractive attribute in some fruits. Triploid bananas and watermelons are intentionally bred because they produce no seeds (and are parthenocarpic).

Heterosis

Hybrids are sometimes stronger than either parent variety, a phenomenon most common with plant hybrids, which when present is known as *hybrid vigor* (heterosis) or heterozygote advantage. A transgressive phenotype is a phenotype displaying more extreme characteristics than either of the parent lines. Plant breeders use several techniques to produce hybrids, including line breeding and the formation of complex hybrids. An economically important example is hybrid maize (corn), which provides a considerable seed yield advantage over open pollinated varieties. Hybrid seed dominates the commercial maize seed market in the United States, Canada and many other major maize-producing countries.

Examples of Plant Hybrids

The multiplication symbol × (not italicised) indicates a hybrid in the Latin binomial nomenclature. Placed before the binomial it indicates a hybrid between species from different genera (intergeneric hybrid):-

- × *Fatshedera lizei*, a hybrid between *Hedera helix* and *Fatsia japonica*
- × *Heucherella*, a hybrid genus between *Heuchera* and *Tiarella*
- × *Philageria veitchii* is a hybrid between *Lapageria rosea* and *Philesia magellanica*; it is more similar in appearance to the former
- Leyland cypress, × *Cupressocyparis leylandii* hybrid between Monterey cypress and Nootka cypress
- Triticale, × *Triticosecale* a wheat–rye hybrid
- × *Urceocharis*, a hybrid between *Eucharis* and *Urceolina*

Interspecific plant hybrids include:

- *Dianthus* × *allwoodii* (*Dianthus caryophyllus* × *Dianthus plumarius*)
- Limequat *Citrus* × *floridana*, key lime *Citrus aurantiifolia* and kumquat *Citrus japonica* hybrid
- Loganberry *Rubus* × *loganobaccus*, a hybrid between raspberry *Rubus idaeus* and blackberry *Rubus ursinus*
- London plane (*Platanus orientalis* × *Platanus occidentalis*), thus forming *Platanus* × *acerifolia*
- *Magnolia* × *alba* (*Magnolia champaca* × *Magnolia montana*)
- Peppermint, a hybrid between spearmint and water mint
- *Quercus* × *warei* (*Quercus robur* × *Quercus bicolor*) 'Nadler' (marketed in the

United States under the trade name Kindred Spirit hybrid oak)

- Tangelo, a hybrid of a Mandarin orange and a pomelo which may have been developed in Asia about 3,500 years ago
- Wheat; most modern and ancient wheat breeds are themselves hybrids. Bread wheat is a hexaploid hybrid of three wild grasses; durum (pasta) wheat is a tetraploid hybrid of two wild grasses
- Grapefruit, hybrid between a pomelo and the Jamaican sweet orange

Some natural hybrids:

- *Iris albicans*, a sterile hybrid which spreads by rhizome division
- Evening primrose, a flower which was the subject of famous experiments by Hugo de Vries on polyploidy and diploidy

Hybrids in Nature

Hybridization between two closely related species is actually a common occurrence in nature but is also being greatly influenced by anthropogenic changes as well. Hybridization is a naturally occurring genetic process where individuals from two genetically distinct populations mate. It can occur both intraspecifically, between different distinct populations within the same species, and interspecifically, between two different species. Hybrids can be either sterile/not viable or viable/fertile. This affects the impact which the hybrid will have on its and other populations that it interacts with. Many hybrid zones are known where the ranges of two species meet, and hybrids are continually produced in great numbers. These hybrid zones are useful as biological model systems for studying the mechanisms of speciation (Hybrid speciation). Recently DNA analysis of a bear shot by a hunter in the North West Territories confirmed the existence of naturally-occurring and fertile grizzly–polar bear hybrids.

Anthropogenic Hybridization

Changes to the environment caused by humans, such as fragmentation and Introduced species, are becoming more widespread. This increases the challenges in managing certain populations that are experiencing introgression, and is a focus of conservation genetics.

Introduced Species and Habitat Fragmentation

Humans have introduced species worldwide to environments for a long time, both intentionally such as establishing a population to be used as a biological control, and unintentionally such as accidental escapes of individuals out of agriculture. This causes drastic global effects on various populations, including through hybridization.

When habitats become broken apart, one of two things can occur, genetically speaking. The first is that populations that were once connected can be cut off from one another, preventing their genes from interacting. Occasionally, this will result in a population of one species breeding with a population of another species as a means of surviving such as the case with the red wolves. Their population numbers being so small, they needed another means of survival. Habitat fragmentation also led to the influx of generalist species into areas where they would not have been, leading to competition and in some cases interbreeding/incorporation of a population into another. In this way, habitat fragmentation is essentially an indirect method of introducing species to an area.

The Hybridization Continuum

There is a kind of continuum with three semi-distinct categories dealing with anthropogenic hybridization: hybridization without Introgression, hybridization with widespread introgression, and essentially a Hybrid swarm. Depending on where a population falls along this continuum, the management plans for that population will change. Hybridization is currently an area of great discussion within Wildlife management and habitat management. Global climate change is creating other changes such as difference in population distributions which are indirect causes for an increase in anthropogenic hybridization.

Consequences

Hybridization can be a less discussed way toward extinction than within detection of where a population lies along the hybrid continuum. The dispute of hybridization is how to manage the resulting hybrids. When a population experiences hybridization with substantial introgression, there still exists parent types of each set of individuals. When a complete hybrid swarm is created, all the individuals are hybrids.

Management of Hybrids

Conservationists disagree on when is the proper time to give up on a population that is becoming a hybrid swarm or to try and save the still existing pure individuals. Once it becomes a complete mixture, we should look to conserve those hybrids to avoid their loss. Most leave it as a case-by-case basis, depending on detecting of hybrids within the group. It is nearly impossible to regulate hybridization via policy because hybridization can occur beneficially when it occurs "naturally" and there is the matter of protecting those previously mentioned hybrid swarms because if they are the only remaining evidence of prior species, they need to be conserved as well.

Expression of Parental Traits in Hybrids

When two distinct types of organisms breed with each other, the resulting hybrids typically have intermediate traits (e.g., one parent has red flowers, the other has white, and the hybrid, pink flowers). Commonly, hybrids also combine traits seen only separately in one parent or the other (e.g., a bird hybrid might combine the yellow head of one parent with the orange belly of the other).

In a hybrid, any trait that falls outside the range of parental variation is termed heterotic. Heterotic hybrids do have new traits, that is, they are not intermediate. *Positive heterosis* produces more robust hybrids, they might be stronger or bigger; while the term *negative heterosis* refers to weaker or smaller hybrids. Heterosis is common in both animal and plant hybrids. For example, hybrids between a lion and a tigress ("ligers") are much larger than either of the two progenitors, while a tigon (lioness × tiger) is smaller. Also the hybrids between the common pheasant (*Phasianus colchicus*) and domestic fowl (*Gallus gallus*) are larger than either of their parents, as are those produced between the common pheasant and hen golden pheasant (*Chrysolophus pictus*). Spurs are absent in hybrids of the former type, although present in both parents.

Genetic Mixing and Extinction

Regionally developed ecotypes can be threatened with extinction when new alleles or genes are introduced that alter that ecotype. This is sometimes called genetic mixing. Hybridization and introgression of new genetic material can lead to the replacement of local genotypes if the hybrids are more fit and have breeding advantages over the indigenous ecotype or species. These hybridization events can result from the introduction of non-native genotypes by humans or through habitat modification, bringing previously isolated species into contact. Genetic mixing can be especially detrimental for rare species in isolated habitats, ultimately affecting the population to such a degree that none of the originally genetically distinct population remains.

Effect on Biodiversity and Food Security

In agriculture and animal husbandry, the Green Revolution's use of conventional hybridization increased yields by breeding "high-yielding varieties". The replacement of locally indigenous breeds, compounded with unintentional cross-pollination and crossbreeding (genetic mixing), has reduced the gene pools of various wild and indigenous breeds resulting in the loss of genetic diversity. Since the indigenous breeds are often well-adapted to local extremes in climate and have immunity to local pathogens, this can be a significant genetic erosion of the gene pool for future breeding. Therefore, commercial plant geneticists strive to breed "widely adapted" cultivars to counteract this tendency.

Limiting Factors

A number of conditions exist that limit the success of hybridization; the most obvious is great genetic diversity between most species. But in animals and plants that are more closely related, hybridization barriers can include morphological differences, differing times of fertility, mating behaviors and cues, physiological rejection of sperm cells or the developing embryo.

In plants, barriers to hybridization include blooming period differences, different pollinator vectors, inhibition of pollen tube growth, somatoplastic sterility, cytoplasmic-genic male sterility and structural differences of the chromosomes.

Mythical, Legendary and Religious Hybrids

Ancient folktales often contain mythological creatures, sometimes described as hybrids (e.g., hippogriff as the offspring of a griffin and a horse, and Minotaur, the offspring of Pasiphaë and a white bull). More often they are kind of chimera, i.e., a composite of the physical attributes of two or more kinds of animals, mythical beasts, and often humans, with no suggestion that they are the result of interbreeding, e.g., harpies, mermaids, and centaurs.

In the Bible, the Old Testament contains several passages which talk about a first generation of hybrid giants who were known as the Nephilim. The Book of Genesis (6:4) states that "the sons of God went to the daughters of humans and had children by them". As a result, the offspring were born as hybrid giants who became mighty heroes of old and legendary famous figures of ancient times. In addition, the Book of Numbers (13:33) says that the descendants of Anak came from the Nephilim, whose bodies looked exactly like men, but with an enormous height. According to the apocryphal Book of Enoch the Nephilim were wicked sons of fallen angels who had lusted with attractive women.

PUREBRED

Purebreds, also called purebreeds, are cultivated varieties or *cultivars* of an animal species, achieved through the process of selective breeding. When the lineage of a purebred animal is recorded, that animal is said to be *pedigreed.*

The term *purebred* is occasionally confused with the proper noun *Thoroughbred,* which refers exclusively to a specific breed of horse, one of the first breeds for which a written national stud book was created since the 18th century. Thus a purebred animal should never be called a "thoroughbred" unless the animal actually is a registered Thoroughbred horse.

True Breeding

In the world of selective animal breeding, to "breed true" means that specimens of an animal breed will breed true-to-type when mated like-to-like; that is, that the progeny of any two individuals in the same breed will show consistent, replicable and predictable characteristics. A puppy from two purebred dogs of the same breed, for example, will exhibit the traits of its parents, and not the traits of all breeds in the subject breed's ancestry.

However, breeding from too small a gene pool, especially direct inbreeding, can lead to the passing on of undesirable characteristics or even a collapse of a breed population due to inbreeding depression. Therefore, there is a question, and often heated controversy, as to when or if a breed may need to allow "outside" stock in for the purpose of improving the overall health and vigor of the breed.

Because pure-breeding creates a limited gene pool, purebred animal breeds are also susceptible to a wide range of congenital health problems. This problem is especially prevalent in competitive dog breeding and dog show circles due to the singular emphasis on aesthetics rather than health or function. Such problems also occur within certain segments of the horse industry for similar reasons. The problem is further compounded when breeders practice inbreeding. The opposite effect to that of the restricted gene pool caused by pure-breeding is known as hybrid vigor, which generally results in healthier animals.

Pedigrees

A purebred Arabian horse.

A **pedigreed animal** is one that has its ancestry recorded. Often this is tracked by a major registry. The number of generations required varies from breed to breed, but all pedigreed animals have papers from the registering body that attest to their ancestry.

The word «pedigree» appeared in the English language in 1410 as «pee de Grewe», «pedegrewe» or «pedegru», each of those words being borrowed to the Middle French «pié de grue», meaning «crane foot». This comes from a visual analogy between the trace of the bird›s foot and the three lines used in the English official registers to show the ramifications of a genealogical tree.

Sometimes the word *purebred* is used synonymously with *pedigreed*, but purebred refers to the animal having a known ancestry, and pedigree refers to the written record of breeding. Not all purebred animals have their lineage in written form. For example, until the 20th century, the Bedouin people of the Arabian peninsula only recorded the ancestry of their Arabian horses via an oral tradition, supported by the swearing of religiously based oaths as to the asil or "pure" breeding of the animal. Conversely, some animals may have a recorded pedigree or even a registry, but not be considered "purebred". Today the modern Anglo-Arabian horse, a cross of Thoroughbred and Arabian bloodlines, is considered such a case.

Purebreds by Animal

Purebred Dogs

Pure-breed Zwergschnauzer

A *purebred dog* is a dog of a modern breed of dog, with written documentation showing the individual purebred dog's descent from its breeds' foundation stock. In dogs, the term breed is used two ways: loosely, to refer to dog types or landraces of dog (also called natural breeds or ancient breeds); or more precisely, to refer to modern breeds of dog, which are documented so as to be known to be descended from specific ancestors, that closely resemble others of their breed in appearance, movement, way of working and other characters; and that reproduce with offspring closely resembling each other and their parents. Purebred dogs are breeds in the second sense.

New breeds of dog are constantly being created, and there are many websites for new breed associations and breed clubs offering legitimate registrations for new or rare breeds. When dogs of a new breed are «visibly similar in most characteristics» and have reliable documented descent from a «known and designated foundation stock», then they can then be considered members of a breed, and, if an individual dog is documented and registered, it can be called *purebred*.

Purebred Horses

The domestication of the horse resulted in a small number of domesticated stallions (possibly a single male ancestor) being crossed on wild mares that had adapted to local conditions. This ultimately produced horses of four basic body types, once thought to be wild prototypes, but now considered to be landraces. Many of these animals were then bred true to original type by selected breeding, though emphasizing certain inherent traits (such as a good temperament, suitable to training by humans) to a greater degree than others. In other cases, horses of different body types were cross bred until a desired characteristic was achieved and bred true.

Written and oral histories of various animals or pedigrees of certain types of horse have been kept throughout history, though breed registry stud books trace only to about the 13th century, at least in Europe, when pedigrees were tracked in writing, and the practice of declaring a type of horse to be a breed or a purebred became more widespread.

Certain horse breeds, such as the Andalusian horse and the Arabian horse, are claimed by aficionados of the respective breeds to be ancient, near-pure descendants from an ancient wild prototype, though mapping of the horse genome as well as the mtDNA and y-DNA of various breeds has largely disproved such claims.

Purebred Cats

Purebred Himalayan

A cat whose ancestry is formally registered is called a pedigreed or purebred cat. Technically, a *purebred* cat is one whose ancestry contains only individuals of the same breed. A *pedigreed* cat is one whose ancestry is recorded, but may have ancestors of different breeds.

The list of cat breeds is quite large: most cat registries actually recognize between 30 and 40 breeds of cats, and several more are in development, with one or more new breeds being recognized each year on average, having distinct features and heritage. Owners and breeders compete in cat shows to see whose animal bears the closest resemblance (best *conformance*) to an idealized definition, based on breed type and the breed standard for each breed.

Because of common crossbreeding in populated areas, many cats are simply identified as belonging to the mixed types of domestic long-haired and domestic short-haired cat, depending on their type of fur.

Some original cat breeds that have a distinct phenotype that is the main type occurring naturally as the dominant domesticated cat type in their region of origin are sometimes considered subspecies and in the past received names as such, although this is no longer supported by feline biologists. with *F. silvestris* the wild cat and *F. s. catus* the domestic. Some of these cat breeds (with their invalid scientific names for historical interest) are:

- *F. catus anura* - the Manx
- *F. catus siamensis* - the Siamese
- *F. catus cartusenensis* - the Chartreux
- *F. catus angorensis* - the Turkish Angora

Purebred Livestock

Purebred Barbados Blackbelly hair sheep.

Most domesticated farm animals also have true-breeding breeds and breed registries, particularly cattle, sheep, goats, rabbits, and pigs. While animals bred strictly for market sale are not always purebreds, or if purebred may not be registered, most livestock producers value the presence of purebred genetic stock for the consistency of traits such animals provide. It is common for a farm's male breeding stock in particular to be of purebred, pedigreed lines.

In cattle, some breeders associations make a difference between "purebred" and "full blood". Full blood cattle are fully pedigreed animals, where every ancestor is registered in the herdbook and shows the typical characteristics of the breed. Purebred are those animals that have been bred-up to purebred status as a result of using full blood animals to cross with an animal of another breed. The breeders association rules the percentage of fullblood genetics required for an animal to be considered purebred, usually above 87.5%.

Charolais calves that were transferred as embryos, with their Angus and Hereford recipient mothers.

Artificial breeding via artificial insemination or embryo transfer is often used in sheep and cattle breeding to quickly expand, or improve purebred herds. Embryo transfer techniques allow top quality female livestock to have a greater influence on the genetic advancement of a herd or flock in much the same way that artificial insemination has allowed greater use of superior sires.

Wild Species, Landraces, and Purebred Species

Breeders of purebred domesticated species discourage crossbreeding with wild species, unless a deliberate decision is made to incorporate a trait of a wild ancestor back into a given breed or strain. Wild populations of animals and plants have evolved naturally over millions of years through a process of natural selection in contrast to human controlled Selective breeding or artificial selection for desirable traits from the human point of view. Normally, these two methods of reproduction operate independently of one another. However, an intermediate form of selective breeding, wherein animals or plants are bred by humans, but with an eye to adaptation to natural region-specific conditions and an acceptance of natural selection to weed out undesirable traits, created many ancient domesticated breeds or types now known as landraces.

Many times, domesticated species live in or near areas which also still hold naturally evolved, region-specific wild ancestor species and subspecies. In some cases, a domesticated species of plant or animal may become feral, living wild. Other times, a wild species will come into an area inhabited by a domesticated species. Some of these situations lead to the creation of hybridized plants or animals, a cross between the native species and a domesticated one. This type of crossbreeding, termed genetic pollution by those who are concerned about preserving the genetic base of the wild species, has become a major concern. Hybridization is also a concern to the breeders of purebred species as well, particularly if the gene pool is small and if such crossbreeding or hybridization threatens the genetic base of the domesticated purebred population.

The concern with genetic pollution of a wild population is that hybridized animals and plants may not be as genetically strong as naturally evolved region specific wild ancestors wildlife which can survive without human husbandry and have high immunity to natural diseases. The concern of purebred breeders with wildlife hybridizing a domesticated species is that it can coarsen or degrade the specific qualities of a breed developed for a specific purpose, sometimes over many generations. Thus, both purebred breeders and wildlife biologists share a common interest in preventing accidental hybridization.

REFERENCES

1. "Hybrid". Merriam Webster. Retrieved 12 June 2014.
2. Rieger, R.; Michaelis A.; Green, M. M. (1991). Glossary of Genetics (5th ed.). Springer-Verlag. ISBN 0-387-52054-6, p. 256
3. Keeton, William T. 1980. Biological science. New York: Norton. ISBN 0-393-95021-2 p. A9
4. Ghigi A. 1936. "Galline di faraone e tacchini" Milano (Ulrico Hoepli)
5. askoxford.com
6. Oxford English Dictionary Online, Oxford University Press 2007
7. "When the sire is a lion the result is termed a Liger, whilst the converse is a Tigon." Edward George Boulenger, World natural history, B. T. Batsford ltd., 1937, p. 40
8. Wricke, Gunter, and Eberhard Weber. 1986. Quantitative genetics and selection in plant breeding. Berlin: W. de Gruyter, p. 257
9. J. O. Rawlings, C. Clark Cockerham Analysis of Double Cross Hybrid Populations. J. O. Rawlings, C. Clark Cockerham Biometrics, Vol. 18, No. 2 (June 1962), pp. 229-244 doi:10.2307/2527461
10. Roy, Darbeshwar. 2000. Plant breeding analysis and exploitation of variation. Pangbourne, UK: Alpha Science International, p. 446
11. Toogood, A.(ed.) (1999). Plant Propagation (1st American ed.). American Horticultural Society. ISBN 0-7894-5520-X, p. 21
12. Arnold, M.L. (1996). Natural Hybridization and Evolution. New York: Oxford University Press. p. 232. ISBN 978-0-19-509975-1.
13. Keeton, William T. 1980. Biological science. New York: Norton. ISBN 0-393-95021-2, p. 800
14. Rong, R; Chandley, A. C.; Song, J; McBeath, S; Tan, P. P.; Bai, Q; Speed, R. M. (1988). "A fertile mule and hinny in China". Cytogenetics and cell genetics. 47 (3): 134–9. doi:10.1159/000132531. PMID 3378453.
15. Esch, Mary. "Study: Eastern wolves are hybrids with coyotes". Associated Press.

16. Rutledge, Linda Y.; et. al. (2012). "Conservation genomics in perspective: A holistic approach to understanding Canis evolution in North America" (PDF). Biological Conservation. 155: 186–192. doi:10.1016/j.biocon.2012.05.017. Retrieved 1 July 2013.
17. Chambers, Steven M.; et. al. (2012). "An account of the taxonomy of North American wolves from morphological and genetic analyses". North American Fauna. 77: 1–67. doi:10.3996/nafa.77.0001. Retrieved 2 July 2013.
18. Dumbacher, J., Review of Proposed Rule Regarding Status of the Wolf Under the Endangered Species Act, NCEAS (January 2014)
19. Frost, Grant, et. al. The amphibian tree of life. Bulletin of the American Museum of Natural History. No. 297. New York. Issued 15 March 2006.
20. Soubrier, Julien; et. al. (18 October 2016). "Early cave art and ancient DNA record the origin of European bison". Nature Communications. 7. doi:10.1038/ncomms13158. ISSN 2041-1723.
21. Daley, Jason. "Cave Paintings Help Unravel the Mystery of the 'Higgs Bison'". Smithsonian. Retrieved 20 October 2016.
22. R.W. Bulliet The Camel and the Wheel (Cambridge Mass. 1975), pp. 164-75
23. "Saving Wild Places"
24. Godzilla vs. Godzilla—How the Chinese Giant Salamander is taking a toll on its Japanese Comic Counterpart Amphibians.org (May 2014)
25. Voloder, Dubravka (3 January 1012). "Print Email Facebook Twitter More World-first hybrid sharks found off Australia". ABC News. Retrieved 5 January 2012.
26. Grula, John W.; Taylor, Orley R. (1980). "The Effect of X-Chromosome Inheritance on Mate-Selection Behavior in the Sulfur Butterflies, Colias eurytheme and C. Philodice". Evolution. 34 (4): 688–95. doi:10.2307/2408022.
27. "Lilium Hybrids". Pacific Bulb Society. Retrieved 22 March 2015.
28. Warschefsky, E.; Penmetsa, R. V.; Cook, D. R.; von Wettberg, E. J. B. (8 October 2014). "Back to the wilds: Tapping evolutionary adaptations for resilient crops through systematic hybridization with crop wild relatives". American Journal of Botany. 101 (10): 1791–1800. doi:10.3732/ajb.1400116. PMID 25326621.
29. University of Colorado Principles of Genetics (MCDB 2150) Lecture 33: Chromosomal changes: Monosomy, Trisomy, Polyploidy, Structural Changes Archived 14 October 2012 at the Wayback Machine.
30. "Evaluating the utility of Arabidopsis thaliana as a model for understanding heterosis in hybrid crops", Euphytica, Springer Netherlands, 156 (1-2): 157–171, July 2007, doi:10.1007/s10681-007-9362-1, ISSN 0014-2336
31. Rieseberg, Loren H.; Margaret A. Archer; Robert K. Wayne (July 1999). "Transgressive segregation, adaptation and speciation". Heredity. 83 (4): 363–372. doi:10.1038/sj.hdy.6886170. PMID 10583537.
32. Smith C. Wayne. Corn: Origin, History, Technology, and Production. Wiley Series in Crop Science, 2004, p. 332.

33. Allendorf, Fred W.; R.F. Leary; P. Spruell; J.K. Wenburg (November 2001). "The problems with hybrids: setting conservation guidelines". TRENDS in Ecology & Evolution. 16 (11): 613–622. doi:10.1016/S0169-5347(01)02290-X.
34. Allendorf, Fred (2007). Conservation and the Genetics of Populations. Malden, MA: Blackwell Publishing. p. 534.
35. Allendorf, Fred (2007). Conservation and the Genetics of Populations. Malden, MA: Blackwell Publishing. pp. 421–448.
36. "Hybrid bear shot dead in Canada". BBC News. 13 May 2006.
37. Ehrlich, Paul; John Holdren (26 March 1971). "Impact of population Growth". Science. 171 (3977): 1212–1216. doi:10.1126/science.171.3977.1212.
38. Vitousek, Peter; et. al. (1997). "Introduced Species: A Significant Component of Human-cause Global Change". New Zealand Journal of Ecology. 21 (1): 1–16.
39. McCarthy, Eugene M. 2006. Handbook of Avian Hybrids of the World. Oxford: Oxford University Press, pp. 16-17
40. M 2006. Handbook of Avian Hybrids of the World. Oxford: Oxford University Press. P. 17.
41. Darwin, C. 1868. Variation of Animals and Plants under Domestication, vol. II, p. 125
42. Spicer, J. W. G. 1854. Note on hybrid gallinaceous birds. The Zoologist, 12: 4294-4296 (see p. 4295).
43. Mooney, H. A.; Cleland, E. E. (2001). "The evolutionary impact of invasive species". Proc Natl Acad Sci U S A. 98 (10): 5446–5451. doi:10.1073/pnas.091093398. PMC 33232 . PMID 11344292.
44. Rhymer, J. M.; Simberloff, D. (1996). "Extinction by Hybridization and Introgression". Annual Review of Ecology and Systematics. 27: 83–109. doi:10.1146/annurev.ecolsys.27.1.83.
45. Brad M. Potts, Robert C. Barbour, Andrew B. Hingston (2001) Genetic Pollution from Farm Forestry using eucalypt species and hybrids; A report for the RIRDC/L&WA/FWPRDC; Joint Venture Agroforestry Program; RIRDC Publication No 01/114; RIRDC Project No CPF - 3A; ISBN 0-642-58336-6; ISSN 1440-6845; Australian Government, Rural Industrial Research and Development Corporation
46. "Genetic Pollution: The Great Genetic Scandal" Devinder Sharma Archived 18 May 2009 at the Wayback Machine.; Bulletin 28. hosted by www.farmedia.org
47. Troyer, A. Forrest. Breeding Widely Adapted Cultivars: Examples from Maize. Encyclopedia of Plant and Crop Science, 27 February 2004.
48. "Barriers to hybridization of Solanum bulbocastanumDun. and S. VerrucosumSchlechtd. and structural hybridity in their F1 plants", Euphytica, Springer Netherlands, 25 (1): 1–10, January 1976, doi:10.1007/BF00041523, ISSN 0014-2336
49. "Traditions of the Bible: A Guide to the Bible As It Was at the Start of the Common Era", James L. Kugel (2009), Harvard University Press, p. 198

50. "The Bible as it was", James L. Kugel (1997), , Harvard University Press, p. 110
51. Gregory A. Boyd, God at War: The Bible & Spiritual Conflict, p. 177
52. (2014) Plant Breeding and Genetics Joint FAO/IAEA Division of Nuclear Techniques in Food and Agriculture, Retrieved 31 July 2014
53. Schouten, H. J.; Jacobsen, E. (2007). "Are Mutations in Genetically Modified Plants Dangerous?". Journal of Biomedicine and Biotechnology. 2007: 1. doi:10.1155/2007/82612.
54. M.K. Maluszynsk, K. Nichterlein, L. van Zanten & B.S. Ahloowalia (2000). "Officially released mutant varieties – the FAO/IAEA Database". Mutation Breeding Review (12): 1–84.
55. Ahloowali, B.S. (2004). "Global impact of mutation-derived varieties" (PDF). Euphytica. 135: 187–204. doi:10.1023/b:euph.0000014914.85465.4f. Retrieved 20 April 2011.
56. Kaskey, Jack (21 November 2013) The Scariest Veggies of Them All Bloomberg Business Week, Retrieved 31 July 2014
57. "New Citrus Variety Released by UC Riverside is Very Sweet, Juicy and Low-seeded".
58. Broad, William J. (28 August 2007). "Useful Mutants, Bred With Radiation". New York Times. Retrieved 20 April 2011.
59. Atomic Gardens: Public Perceptions & Public Policy, Life Sciences Foundation Magazine, Spring 2012.
60. Smith, Peter (2011-04-12). "How Radiation is Changing the Foods that You Eat". GOOD. GOOD Worldwide, Inc. Retrieved 2011-07-16.
61. Pathirana, R. Plant Mutation Breeding in Agriculture. CAB Reviews: Perspectives in Agriculture, Veterinary Science, Nutrition and Natural Resources. 2011 6 No 032
62. Johnson, Paige. "Atomic Gardens". Retrieved 20 April 2011.
63. Useful Mutants, Bred With Radiation, by William J. Broad, New York Times, August 28, 2007.
64. Discussion Document Excluded Methods Terminology, National Organic Standards Board GMO ad hoc Subcommittee paper, U.S. Agricultural Marketing Service, published February 6, 2013.
65. Mendel in the Kitchen: A Scientist's View of Genetically Modified Foods, By Nina V. Fedoroff and Nancy Marie Brow, pg. 17, Joseph Henry Press, 2004.
66. Golden Promise Organic Ale
67. Wasatch Organic Rio Red Grapefruit
68. Stoilov, L.; Gecheff, K. (2009). Shu, Q.Y., ed. "Restriction Endonucleases as a Tool for In Vivo Induction of Chromosomal and DNA Damage in Barley Genome" (PDF). Induced Plant Mutations in the Genomics Era – via Food and Agriculture Organization of the United Nations.

69. Liu, L.X.; Guo, H.J.; Zhao, L.S.; Wang, J.; Zhao, S.R. (2009). Shu, Q.Y., ed. "Achievements and Perspectives of Crop Space Breeding in China" (PDF). Induced Plant Mutations in the Genomics Era – via Food and Agriculture Organization of the United Nations.
70. "Mutant Lotus Bred in Space on Display in Chongqing, China".
71. Tanaka, A. (2009). Shu, Q.Y., ed. "Establishment of Ion Beam Technology for Breeding" (PDF). Induced Plant Mutations in the Genomics Era – via Food and Agriculture Organization of the United Nations.
72. Wang, H.; Qiu, S.; Zheng, J.; Jiang, L.; Huang, H.; Huang, Y. (2009). Shu, Q.Y., ed. "Generation of New Rice Cultivars from Mature Pollen Treated with Gamma-Radiation" (PDF). Induced Plant Mutations in the Genomics Era – via Food and Agriculture Organization of the United Nations.
73. Kharkwal, M. C.; Shu, Q. Y. (2008). "The role of induced mutations in world food security" (PDF). Induced plant mutations in the genomics era. Proceedings of an International Joint FAO/IAEA Symposium, 2008 2009. Vienna, Austria: Joint FAO/IAEA Division of Nuclear Techniques in Food and Agriculture, International Atomic Energy Agency: 33–38. ISBN 9789251063248.
74. Shu, Qing-Yao (2012). Forster, B. P.; Nakagawa, H., eds. Plant Mutation Breeding and Biotechnology. CABI Publishing. p. 17. ISBN 978-1780640853.
75. "Lift-off for Chinese space potato". BBC News. 12 February 2007.
76. Ahloowalia, B. S.; Maluszynski, M. (2001). "Production Process in Old and Modern Spring Barley Varieties". Euphytica. 118 (2): 167. doi:10.1023/A:1004162323428.
77. Lipavsky, J. Petr, J. and Hradecká, D, (2002) "Production Process in Old and Modern Spring Barley Varieties" Die Bodenkultur, 53 (1) 2, Page 19
78. Rossi, Luigi (2010). "Il miglioramento genetico del grano duro in Casaccia. Il caso CRESO" Genetic Improvement of Durum Wheat in Casaccia. The Creso Case (PDF). Energia, Ambiente e Innovazione. ENEA. Retrieved 2016-11-29.
79. van Harten, A. M. (1998). Mutation Breeding: Theory and Practical Applications. U. K.: Cambridge University Press. p. 239. ISBN 978-0521470742.
80. Kotobuki, Kazuo. "Japanese pear tree named `Osa Gold`". Retrieved 20 April 2011.
81. (2008) NIAB - Plant Breeding & Genetics Division, Achievements Nuclear Institute for Agriculture and Biology, Faisalabad, Pakistan, Retrieved 16 May 2013
82. (2012) Improved barley varieties - Feeding people from the equator to the arctic Joint FAO/IAEAProgramme, Nuclear Techniques in Food and Agriculture, Retrieved 25 October 2013
83. Forster, B. P. (2001). "Mutation genetics of salt tolerance in barley: An assessment of Golden Promise and other semi-dwarf mutants". Euphytica. 120 (3): 317–328. doi:10.1023/A:1017592618298.
84. Broad, William (2007-08-28). "Useful Mutants, Bred With Radiation". New York Times. Retrieved 2013-06-19.

85. (2012) Successful Mutation Breeding Programmes in Vietnam Joint FAO/IAEAProgramme, Nuclear Techniques in Food and Agriculture, Retrieved 25 October 2013
86. Vinh, M.Q. et al (2009) Current Status and Research Directions of Induced Mutation Application to Seeds Program in Vietnam in Induced Plant Mutations in the Genomics Era, FAO of the UN, Rome, Pp 341-345, Web page version retrieved 25 October 2013
87. (2014) Successful Mutation Breeding Programmes in Vietnam Joint FAO/IAEA Division of Nuclear Techniques in Food and Agriculture, Retrieved 31 July 2014
88. Pathirana, Ranjith (September 6, 2011) Plant mutation breeding in agriculture CAB Reviews: Perspectives in Agriculture, Veterinary Science, Nutrition and Natural Resources (CAB International); 20116 (032): 1 – 20; doi:10.1079/PAVSNNR20116032; ISSN 1749-8848; Retrieved August 6, 2014
89. Conner, J. K. (2003). "Artificial Selection: A Powerful Tool for Ecologists". Ecology. 84 (7): 1650–1660. doi:10.1890/0012-9658(2003)0841650:asaptf 2.0.co;2.
90. Darwin
91. Purugganan, M. D.; Fuller, D. Q. (2009). "The nature of selection during plant domestication". Nature. 457 (7231): 843–8. doi:10.1038/nature07895. PMID 19212403.
92. Buffum, Burt C. (2008). Arid Agriculture; A Hand-Book for the Western Farmer and Stockman. Read Books. p. 232. ISBN 978-1-4086-6710-1.
93. Lush, Jay L. (2008). Animal Breeding Plans. Orchard Press. p. 21. ISBN 978-1-4437-8451-1.
94. Wilczynski, J. Z. (1959). "On the Presumed Darwinism of Alberuni Eight Hundred Years before Darwin". Isis. 50 (4): 459. doi:10.1086/348801.
95. "Robert Bakewell (1725–1795)". BBC History. Retrieved 20 July 2012.
96. Darwin, p. 109
97. Darwin, pp. 197–198
98. Grandin, Temple; Johnson, Catherine (2005). Animals in Translation. New York, New York: Scribner. pp. 69–71. ISBN 0-7432-4769-8.
99. Swallow, JG; Garland; Jr (2005). "Selection experiments as a tool in evolutionary and comparative physiology: insights into complex traits—an introduction to the symposium"(PDF). Integr Comp Biol. 45 (3): 387–390. doi:10.1093/icb/45.3.387. PMID 21676784.
100. Garland T, Jr. (2003). Selection experiments: an under-utilized tool in biomechanics and organismal biology. Ch. 3, Vertebrate Biomechanics and Evolution ed. Bels VL, Gasc JP, Casinos A. PDF
101. Garland T, Jr., Rose MR, eds. (2009). Experimental Evolution: Concepts, Methods, and Applications of Selection Experiments. University of California Press, Berkeley, California.

102. Jain, H. K.; Kharkwal, M. C. (2004). Plant breeding - Mendelian to molecular approaches. Boston, London, Dordecht: Kluwer Academic Publishers. ISBN 1-4020-1981-5.

103. Gjedrem, T & Baranski, M. (2009). Selective breeding in Aquaculture: An Introduction. 1st Edition. Springer. ISBN 978-90-481-2772-6

104. Gjedrem, T. (1985). “Improvement of productivity through breeding schemes”. GeoJournal. 10 (3): 233–241. doi:10.1007/BF00462124.

105. Gjedrem, T. (1983). “Genetic variation in quantitative traits and selective breeding in fish and shellfish”. Aquaculture. 33: 51–72. doi:10.1016/0044-8486(83)90386-1.

106. Thodesen, J. R.; Grisdale-Helland, B.; Helland, S. L. J.; Gjerde, B. (1999). “Feed intake, growth and feed utilization of offspring from wild and selected Atlantic salmon (Salmo salar)”. Aquaculture. 180 (3–4): 237–246. doi:10.1016/s0044-8486(99)00204-5.

107. Storset, A.; Strand, C.; Wetten, M.; Kjøglum, S.; Ramstad, A. (2007). “Response to selection for resistance against infectious pancreatic necrosis in Atlantic salmon (Salmo salar L.)”. Aquaculture. 272: S62. doi:10.1016/j.aquaculture.2007.08.011.

108. Donaldson, L. R.; Olson, P. R. (1957). “Development of Rainbow Trout Brood Stock by Selective Breeding”. Transactions of the American Fisheries Society. 85: 93–101. doi:10.1577/1548-8659(1955)85 93:dortbs 2.0.co;2.

109. Kincaid, H. L.; Bridges, W. R.; von Limbach, B. (1977). “Three Generations of Selection for Growth Rate in Fall-Spawning Rainbow Trout”. Transactions of the American Fisheries Society. 106 (6): 621–628. doi:10.1577/1548-8659(1977)106<621:tgosfg>2.0.co;2.

110. Kause, A.; Ritola, O.; Paananen, T.; Wahlroos, H.; Mäntysaari, E. A. (2005). “Genetic trends in growth, sexual maturity and skeletal deformations, and rate of inbreeding in a breeding programme for rainbow trout (Oncorhynchus mykiss)”. Aquaculture. 247: 177–187. doi:10.1016/j.aquaculture.2005.02.023.

111. Okamoto, N.; Tayama, T.; Kawanobe, M.; Fujiki, N.; Yasuda, Y.; Sano, T. (1993). “Resistance of a rainbow trout strain to infectious pancreatic necrosis”. Aquaculture. 117: 71–76. doi:10.1016/0044-8486(93)90124-h.

112. Hershberger, W. K.; Myers, J. M.; Iwamoto, R. N.; McAuley, W. C.; Saxton, A. M. (1990). “Genetic changes in the growth of coho salmon (Oncorhynchus kisutch) in marine net-pens, produced by ten years of selection”. Aquaculture. 85: 187–197. doi:10.1016/0044-8486(90)90018-i.

113. Neira, R.; Díaz, N. F.; Gall, G. A. E.; Gallardo, J. A.; Lhorente, J. P.; Alert, A. (2006). “Genetic improvement in coho salmon (Oncorhynchus kisutch). II: Selection response for early spawning date” (PDF). Aquaculture. 257: 1–8. doi:10.1016/j.aquaculture.2006.03.001.

114. Kirpichnikov, V. S.; Ilyasov, I.; Shart, L. A.; Vikhman, A. A.; Ganchenko, M. V.; Ostashevsky, A. L.; Simonov, V. M.; Tikhonov, G. F.; Tjurin, V. V. (1993). “Selection of Krasnodar common carp (Cyprinus carpio L.) for resistance to dropsy: Principal results and prospects”. Genetics in Aquaculture. p. 7. doi:10.1016/b978-0-444-81527-9.50006-3. ISBN 9780444815279.

115. Babouchkine, Y.P., 1987. La sélection d'une carpe résistant à l'hiver. In: Tiews, K. (Ed.), Proceedings ofWorld Symposium on Selection,Hybridization, and Genetic Engineering in Aquaculture, Bordeaux 27–30 May 1986, vol. 1. HeenemannVerlagsgesellschaft mbH, Berlin, pp. 447–454.

116. Mai Thien Tran; Cong Thang Nguyen (1993). "Selection of common carp (Cyprinus carpio L.) in Vietnam". Aquaculture. 111: 301–302. doi:10.1016/0044-8486(93)90064-6.

117. Moav, R; Wohlfarth, G (1976). "Two-way selection for growth rate in the common carp (Cyprinus carpio L.)". Genetics. 82 (1): 83–101. PMC 1213447 . PMID 1248737.

118. Bondari, K. (1983). "Response to bidirectional selection for body weight in channel catfish". Aquaculture. 33: 73–81. doi:10.1016/0044-8486(83)90387-3.

119. Langdon, C.; Evans, F.; Jacobson, D.; Blouin, M. (2003). "Yields of cultured Pacific oysters Crassostrea gigas Thunberg improved after one generation of selection". Aquaculture. 220: 227–244. doi:10.1016/s0044-8486(02)00621-x.

120. Nell, J. A.; Sheridan, A. K.; Smith, I. R. (1996). "Progress in a Sydney rock oyster, Saccostrea commercialis (Iredale and Roughley), breeding program". Aquaculture. 144 (4): 295–302. doi:10.1016/0044-8486(96)01328-2.

121. Nell, J. A.; Smith, I. R.; Sheridan, A. K. (1999). "Third generation evaluation of Sydney rock oyster Saccostrea commercialis (Iredale and Roughley) breeding lines". Aquaculture. 170 (3–4): 195–203. doi:10.1016/s0044-8486(98)00408-6.

122. Culloty, S. C.; Cronin, M. A.; Mulcahy, M. I. F. (2001). "An investigation into the relative resistance of Irish flat oysters Ostrea edulis L. to the parasite Bonamia ostreae". Aquaculture. 199 (3–4): 229–244. doi:10.1016/s0044-8486(01)00569-5.

123. Culloty, S. C.; Cronin, M. A.; Mulcahy, M. F. (2004). "Potential resistance of a number of populations of the oyster Ostrea edulis to the parasite Bonamia ostreae". Aquaculture. 237: 41–58. doi:10.1016/j.aquaculture.2004.04.007.

124. Ragone Calvo, L. M.; Calvo, G. W.; Burreson, E. M. (2003). "Dual disease resistance in a selectively bred eastern oyster, Crassostrea virginica, strain tested in Chesapeake Bay". Aquaculture. 220: 69–87. doi:10.1016/s0044-8486(02)00399-x.

125. Goyard, E.; Patrois, J.; Reignon, J.-M.; Vanaa, V.; Dufour, R; Be (1999). "IFREMER's shrimp genetics program". Global Aquaculture Advocate. 2 (6): 26–28.

126. Hetzel, D. J. S.; Crocos, P. J.; Davis, G. P.; Moore, S. S.; Preston, N. C. (2000). "Response to selection and heritability for growth in the Kuruma prawn, Penaeus japonicus". Aquaculture. 181 (3–4): 215–223. doi:10.1016/S0044-8486(99)00237-9.

127. Argue, B. J.; Arce, S. M.; Lotz, J. M.; Moss, S. M. (2002). "Selective breeding of Pacific white shrimp (Litopenaeus vannamei) for growth and resistance to Taura Syndrome Virus". Aquaculture. 204 (3–4): 447–460. doi:10.1016/s0044-8486(01)00830-4.

128. Cock, J.; Gitterle, T.; Salazar, M.; Rye, M. (2009). "Breeding for disease resistance of Penaeid shrimps". Aquaculture. 286: 1–11. doi:10.1016/j.aquaculture.2008.09.011.

129. Tang, K. F. J.; Durand, S. V.; White, B. L.; Redman, R. M.; Pantoja, C. R.; Lightner, D. V. (2000). "Postlarvae and juveniles of a selected line of Penaeus stylirostris are resistant to infectious hypodermal and hematopoietic necrosis virus infection". Aquaculture. 190 (3–4): 203–210. doi:10.1016/s0044-8486(00)00407-5.

130. Inbreeding at Encyclopædia Britannica

131. Nabulsi MM, Tamim H, Sabbagh M, Obeid MY, Yunis KA, Bitar FF (2003). "Parental consanguinity and congenital heart malformations in a developing country". American Journal of Medical Genetics Part A. 116A (4): 342–7. doi:10.1002/ajmg.a.10020. PMID 12522788.

132. Jiménez JA, Hughes KA, Alaks G, Graham L, Lacy RC (1994). "An experimental study of inbreeding depression in a natural habitat" (PDF). Science. 266 (5183): 271–3. doi:10.1126/science.7939661. PMID 7939661.

133. Chen X. (1993). "Comparison of inbreeding and outbreeding in hermaphroditic Arianta arbustorum (L.) (land snail)". Heredity. 71 (5): 456–461. doi:10.1038/hdy.1993.163.

134. Bernstein H, Byerly HC, Hopf FA, Michod RE (1985). "Genetic damage, mutation, and the evolution of sex". Science. 229 (4719): 1277–81. doi:10.1126/science.3898363. PMID 3898363.

135. Michod RE. Eros and Evolution: A Natural Philosophy of Sex. (1994) Perseus Books, ISBN 0-201-40754-X

136. Lynch, Michael. (1991). The Genetic Interpretation of Inbreeding Depression and Outbreeding Depression. Oregon: Society for the Study of Evolution.

137. Livingstone, F. B. (1969). "Genetics, Ecology, and the Origins of Incest and Exogamy". Current Anthropology. 10: 45–62. doi:10.1086/201009.

138. Thornhill, Nancy Wilmsen (1993). The Natural History of Inbreeding and Outbreeding: Theoretical and Empirical Perspectives. Chicago: University of Chicago Press. ISBN 0-226-79854-2.

139. Shields, W. M. 1982. Philopatry, Inbreeding, and the Evolution of Sex. Print. 50–69.

140. Meagher, Shawn; Et (2000). "Male–male competition magnifies inbreeding depression in wild house mice". PNAS. 97: 3324–3329. doi:10.1073/pnas.97.7.3324.

141. Swindell, William R.; et al. (2006). "Selection and Inbreeding Depression: Effects of Inbreeding Rate and Inbreeding Environment". Evolution. 60: 1014–1022. doi:10.1554/05-493.1.

142. Lieberman, D.; Tooby, J.; Cosmides, L. (2003). "Does morality have a biological basis? An empirical test of the factors governing moral sentiments relating to incest". Proceedings of the Royal Society B: Biological Sciences. 270 (1517): 819–826. doi:10.1098/rspb.2002.2290.

143. Pusey, A.; Wolf, M. (1996). "Inbreeding avoidance in animals". Trends in Ecology and Evolution. 11: 201–206. doi:10.1016/0169-5347(96)10028-8. PMID 21237809.

144. Shields, William M. (1982). Philopatry, inbreeding, and the evolution of sex. Albany: State University of New York Press. ISBN 0-87395-618-4.

145. Joly, E (9 December 2011). "The existence of species rests on a metastable equilibrium between inbreeding and outbreeding. An essay on the close relationship between speciation, inbreeding and recessive mutations.". Biology direct. 6: 62. doi:10.1186/1745-6150-6-62. PMC 3275546 . PMID 22152499.

146. Hartl, D.L., Jones, E.W. (2000) Genetics: Analysis of Genes and Genomes. Fifth Edition. Jones and Bartlett Publishers Inc., pp. 105–106. ISBN 0-7637-1511-5.

147. Kingston H M (2002). ABC of Clinical Genetics (3rd ed.). London: BMJ Books. p. 7. ISBN 0-7279-1627-0. PMC 1836181 . PMID 2497870.

148. Wolf, Arthur P.; Durham, William H., eds. (2005). Inbreeding, incest, and the incest taboo: the state of knowledge at the turn. Stanford University Press. p. 3. ISBN 0-8047-5141-2.

149. Griffiths, Anthony J. F.; Jeffrey H. Miller; David T. Suzuki; Richard C. Lewontin; William M. Gelbart (1999). An introduction to genetic analysis. New York: W. H. Freeman. pp. 726–727. ISBN 0-7167-3771-X.

150. Bittles, A. H.; Black, M. L. (2010). "Consanguinity, human evolution, and complex diseases". Proceedings of the National Academy of Sciences. 107 (suppl 1): 1779–1786. doi:10.1073/pnas.0906079106. ISSN 0027-8424. PMC 2868287 . PMID 19805052.

151. Fareed, M.; Afzal, M. (2014). "Evidence of inbreeding depression on height, weight, and body mass index: a population-based child cohort". American Journal of Human Biology. 26 (6): 784–795. doi:10.1002/ajhb.22599. ISSN 1520-6300. PMID 25130378.

152. Fareed, M.; Ahmad, MK.; Anwar, MA.; Afzal, M. (2016). "Impact of consanguineous marriages and degrees of inbreeding on fertility, child mortality, secondary sex ratio, selection intensity and genetic load: a cross-sectional study from Northern India". Pediatric Research. doi:10.1038/pr.2016.177. ISSN 0031-3998. PMID 27632780.

153. Fareed, M.; Afzal, M. (2016). "Increased cardiovascular risks associated with familial inbreeding: a population-based study of adolescent cohort". Annals of Epidemiology. 26: 283–92. doi:10.1016/j.annepidem.2016.03.001. ISSN 1047-2797. PMID 27084548.

154. Bittles AH, Grant JC, Shami SA (1993). "Consanguinity as a determinant of reproductive behaviour and mortality in Pakistan". International Journal of Epidemiology. 22 (3): 463–7. doi:10.1093/ije/22.3.463. PMID 8359962.

155. Kirkpatrick M, Jarne P (2000). "The Effects of a Bottleneck on Inbreeding Depression and the Genetic Load". The American Naturalist. 155 (2): 154–167. doi:10.1086/303312. PMID 10686158.

156. Leck, Charles F. (1980). "Establishment of New Population Centers with Changes in Migration Patterns" (PDF). Journal of Field Ornithology. 51 (2): 168–173. JSTOR 4512538.
157. "ADVS 3910 Wild Horses Behavior", College of Agriculture, Utah State University.
158. Wolf, Arthur P.; Durham, William H., eds. (2005). Inbreeding, incest, and the incest taboo: the state of knowledge at the turn. Stanford University Press. p. 6. ISBN 0-8047-5141-2.
159. Freilich, S.; Hoelzel, A.R. and Choudhury, S.R. Genetic diversity and population genetic structure in the South American sea lion (Otaria flavescens), Department of Anthropology and School of Biological & Biomedical Sciences, University of Durham,U.K.
160. Gilbert, D. A.; Packer, C.; Pusey, A. E.; Stephens, J. C.; O'Brien, S. J. (1991-10-01). "Analytical DNA fingerprinting in lions: parentage, genetic diversity, and kinship". The Journal of Heredity. 82 (5): 378–386. ISSN 0022-1503. PMID 1940281.
161. Ramel, C. (1998). "Biodiversity and intraspecific genetic variation". Pure and Applied Chemistry. 70 (11): 2079. doi:10.1351/pac199870112079.
162. Wielebnowski, N (1996). "Reassessing the Relationship Between Juvenile Mortality and Genetic Monomorphism in Captive Cheetahs". Zoo Biology. 15 (4): 353–369. doi:10.1002/(SICI)1098-2361(1996)15:43.0.CO;2-A.
163. Wright, S (1922). "Coefficients of inbreeding and relationship". American Naturalist. 56: 330–338. doi:10.1086/279872.
164. Malecot, G. 1048. Les Mathématiques de l'hérédité. Masson et Cie, Paris.
165. How to compute and inbreeding coefficient (the path method), Braque du Bourbonnais.
166. Knud Christensen, 4.5 Calculation of inbreeding and relationship, the tabular method, in 14. Genetic calculation applets and other programs.
167. Tave, Douglas (1999). Inbreeding and brood stock management. Food and Agriculture Organization of the United Nations. p. 50. ISBN 978-92-5-104340-0.
168. G2036 Culling the Commercial Cow Herd: BIF Fact Sheet, MU Extension. Extension.missouri.edu. Retrieved on 2013-03-05.
169. "Genetic Evaluation Results". Archived from the original on August 27, 2001.
170. S1008: Genetic Selection and Crossbreeding to Enhance Reproduction and Survival of Dairy Cattle (S-284). Nimss.umd.edu. Retrieved on 2013-03-05.
171. Vogt, Dale; Swartz, Helen A.; Massey, John (October 1993). "Inbreeding: Its Meaning, Uses and Effects on Farm Animals". MU Extension. University of Missouri. Retrieved April 30, 2011.
172. Top Cat Breeds for 2004. Petplace.com. Retrieved on 2013-03-05.
173. Preserving Quality and Genetic Diversity in a Dog Breed. bulldoginformation.com
174. Woodley, Michael A. (2009-05-01). "Inbreeding depression and IQ in a study

of 72 countries". Intelligence. 37 (3): 268–276. doi:10.1016/j.intell.2008.10.007.

175. Kamin, Leon J. (1980-01-01). "Inbreeding depression and IQ.". Psychological Bulletin. 87 (3): 469–478. doi:10.1037/0033-2909.87.3.469.
176. Tadmouri, Ghazi O.; Nair, Pratibha; Obeid, Tasneem; Ali, Mahmoud T. Al; Khaja, Najib Al; Hamamy, Hanan A. (2009-10-08). "Consanguinity and reproductive health among Arabs". Reproductive Health. 6 (1): 17. doi:10.1186/1742-4755-6-17. ISSN 1742-4755. PMC 2765422 . PMID 19811666.
177. Roberts, D. F. (1967-11-11). "Incest, inbreeding and mental abilities.". Br Med J. 4 (5575): 336–337. doi:10.1136/bmj.4.5575.336. ISSN 0007-1447. PMID 6053617.
178. "Human inbreeding avoidance: Culture in nature". Behavioral and Brain Sciences. 6: 91. doi:10.1017/S0140525X00014850.
179. Bittles, AH (2010). "Consanguinity, Genetic Drift, and Genetic Diseases in Populations with Reduced Numbers of Founders". In Speicher, MR; Motulsky, AG; Antonarakis, SE. Vogel and Motulsky's human genetics problems and approaches (4th, rev. ed.). Berlin: Springer-Verlag. pp. 507–528. ISBN 978-3-540-37654-5.
180. Ober, Carole; Hyslop, Terry; Hauck, Walter W. (1999-01-01). "Inbreeding Effects on Fertility in Humans: Evidence for Reproductive Compensation". The American Journal of Human Genetics. 64 (1): 225–231. doi:10.1086/302198. PMC 1377721 . PMID 9915962.
181. Morton, N. E. (1978-08-01). "Effect of inbreeding on IQ and mental retardation". Proceedings of the National Academy of Sciences. 75 (8): 3906–3908. doi:10.1073/pnas.75.8.3906. ISSN 0027-8424. PMID 279005.
182. Bittles, A. H.; Grant, J. C.; Sullivan, S. G.; Hussain, R. (2002-01-01). "Does inbreeding lead to decreased human fertility?". Annals of Human Biology. 29 (2): 111–130. doi:10.1080/03014460110075657.
183. Ober, C.; Elias, S.; Kostyu, D. D.; Hauck, W. W. (1992). "Decreased fecundability in Hutterite couples sharing HLA-DR". American Journal of Human Genetics. 50 (1): 6–14.
184. Diamond, J. M. (1987). "Causes of death before birth". Nature. 329 (6139): 487–488. doi:10.1038/329487a0. PMID 3657971.
185. Stoltenberg, C.; Magnus, P.; Skrondal, A.; Lie, R. T. (1999). "Consanguinity and recurrence risk of stillbirth and infant death". American Journal of Public Health. 89 (4): 517–523. doi:10.2105/ajph.89.4.517.
186. Khlat, M (1989). "Inbreeding effects on fetal growth in Beirut, Lebanon". American Journal of Physical Anthropology. 80: 481–484. doi:10.1002/ajpa.1330800407.
187. Bener, A.; Dafeeah, E. E.; Samson, N. (2012). "Does consanguinity increase the risk of schizophrenia? Study based on primary health care centre visits". Mental Health in Family Medicine. 9 (4): 241–248.
188. Girotto, Giorgia; Mezzavilla, Massimo; Abdulhadi, Khalid; Vuckovic, Dragana; Vozzi, Diego; Alkowari, Moza Khalifa; Gasparini, Paolo; Badii, Ramin (2014-

01-01). "Consanguinity and Hereditary Hearing Loss in Qatar". Human Heredity. 77 (1–4): 175–182. doi:10.1159/000360475. PMID 25060281.

189. Beeche, Arturo (2009). The Gotha: Still a Continental Royal Family, Vol. 1. Richmond, US: Kensington House Books. pp. 1–13. ISBN 978-0-9771961-7-3.
190. Seawright, Caroline. "Women in Ancient Egypt, Women and Law". thekeep.org.
191. Bevan, E.R. "The House of Ptolomey". uchicago.edu.
192. "The Habsburg Lip", Topics in the History of Genetics and Molecular Biology, Fall 2000. Msu.edu. Retrieved on 2013-03-05.
193. "The Imperial House of Habsburg: Chapter 5". Archived from the original on August 27, 2007. Retrieved September 23, 2007.
194. Alvarez, Gonzalo; Ceballos, Francisco C.; Quinteiro, Celsa (2009). "The Role of Inbreeding in the Extinction of a European Royal Dynasty". PLoS ONE. 4 (4): e5174. doi:10.1371/journal.pone.0005174. PMC 2664480 . PMID 19367331.

CHAPTER 4

Artificial Insemination in Animals

Chapter Contents

Artificial insemination is way of treating infertility. It involves direct insertion of semen into a woman's womb.

INTRODUCTION

Artificial insemination (AI) is the deliberate introduction of sperm into a female's uterus or cervix for the purpose of achieving a pregnancy through in vivo fertilization by means other than sexual intercourse. It is a fertility treatment for humans, and is a common practice in animal breeding, including dairy cattle (see Frozen bovine semen) and pigs.

Artificial insemination may employ assisted reproductive technology, sperm donation and animal husbandry techniques. Artificial insemination techniques available include intracervical insemination and intrauterine insemination. The primary beneficiaries of artificial insemination are women who desire to give birth to their own child. They may be in a heterosexual relationship but with the male partner suffering from male infertility, in a lesbian relationship or are single women. Intracervical insemination (ICI) is the easiest and most common insemination technique and can be used in the home for self-insemination without medical practitioner assistance. Compared to natural insemination (i.e., insemination by sexual intercourse), artificial insemination can be more expensive and more invasive, and may require professional assistance.

There are laws in some countries which restrict and regulate who can donate sperm and who is able to receive artificial insemination, and the consequences of such insemination. Some women who live in a jurisdiction which does not permit artificial insemination in the circumstance in which she finds herself may travel to another jurisdiction which permits it.

IN HUMANS

General

In the case of heterosexual couples where the woman is finding it difficult to conceive, before artificial insemination is turned to as the solution to impregnate a woman, doctors will require an examination of both the male and female involved in order to remove any and all physical hindrances that are preventing them from naturally achieving a pregnancy. The couple is also given a fertility test to determine the motility, number, and viability of the male's sperm and the success of the female's ovulation. From these tests, the doctor may or may not recommend a form of artificial insemination.

The sperm used in artificial insemination may be provided by either the woman's husband or partner (partner sperm) or by a known or anonymous sperm donation (donor sperm). Partner sperm may be used when a male partner's physical limitation impedes his ability to impregnate her by sexual intercourse, or the partner's sperm had been frozen in anticipation of some medical procedure or if the partner has since died. In other cases, sperm from an anonymous or known donor may be used.

Though there may be legal, religious and cultural differences in these and other characterizations, the manner in which the sperm is actually used in AI would be the same, If the procedure is successful, the woman will conceive and carry a baby to term in the normal manner. A pregnancy resulting from artificial insemination will be no different from a pregnancy achieved by sexual intercourse. In all cases, the woman would be the biological mother of any child produced by AI, and the male whose sperm is used would be the biological father.

There are multiple methods used to obtain the semen necessary for artificial insemination. Some methods require only men, while others require a combination of a male and female. Those that require only men to obtain semen are masturbation, a rectal massage, involuntary pollution (the collecting of nocturnal emission), or the aspiration of sperm by means of a puncture of the testicle and epididymus. Methods of collecting semen that involve a combination of a male and female include interrupted intercourse, intercourse with a 'collection condom', or the post coital aspiration of the semen from the vagina.

There are a number of reasons why a woman would use artificial insemination to achieve pregnancy. For example, a woman's immune system may be rejecting her partner's sperm as invading molecules. Women who have issues with the cervix, such as cervical scarring, cervical blockage from endometriosis, or thick cervical mucus may also benefit from artificial insemination since the sperm must pass through the cervix to result in fertilization.

Donor sperm is increasingly used where a single woman without a male partner or a lesbian couple wish to have a biological child. A couple where one person is transgender and no longer has gonads may also use donor sperm to become pregnant.

Preparations

Timing is critical, as the window and opportunity for fertilization is little more than twelve hours from the release of the ovum. To increase the chance of success, the woman's menstrual cycle is closely observed, often using ovulation kits, ultrasounds or blood tests, such as basal body temperature tests over, noting the color and texture of the vaginal mucus, and the softness of the nose of her cervix. To improve the success rate of AI, drugs to create a stimulated cycle may be used, but the use of such drugs also results in an increased chance of a multiple birth.

Sperm can be provided fresh or washed. The washing of sperm increases the chances of fertilization. Pre- and post-concentration of motile sperm is counted. Sperm from a sperm bank will be frozen and quarantined for a period and the donor will be tested before and after production of the sample to ensure that he does not carry a transmissible disease. For fresh shipping, a semen extender is used.

If sperm is provided by a private donor, either directly or through a sperm agency, it is usually supplied fresh, not frozen, and it will not be quarantined. Donor sperm provided in this way may be given directly to the recipient woman or her partner, or it may be transported in specially insulated containers. Some donors have their own freezing apparatus to freeze and store their sperm.

Techniques

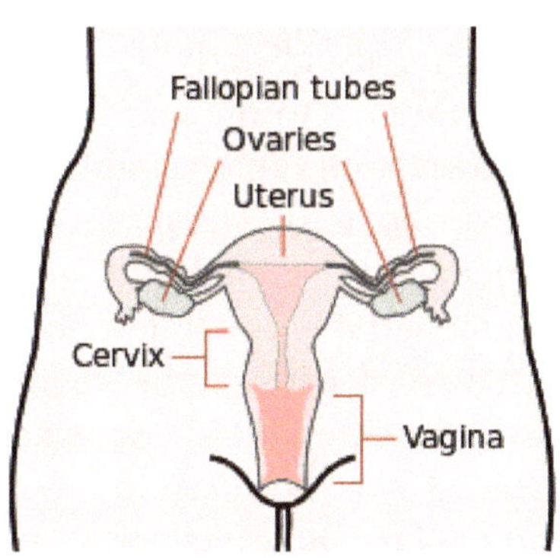

The Human female reproductive system. The cervix is part of the uterus. The cervical canal connects the interiors of the uterus and vagina.

Semen used in insemination is used either fresh, raw or frozen. Where donor sperm is supplied by a sperm bank, it will always be quarantined and frozen and will need to be thawed before use. When an ovum is released, semen is introduced into the woman's vagina, uterus or cervix, depending on the method being used. Sperm is occasionally inserted twice within a 'treatment cycle'.

Intracervical Insemination

Intracervical insemination (ICI) involves injection of unwashed or raw semen into the cervix with a needleless syringe. Where fresh semen is used this must be allowed to liquefy before inserting it into the syringe, or alternatively, the syringe may be back-loaded. After the syringe has been filled with semen, enclosed air is removed by gently pressing the plunger forward before inserting the syringe into the vagina. Care is optimal when inserting the syringe, so that the tip is as close to the entrance to the cervix as possible. A vaginal speculum may be used to hold open the vagina so that the cervix may be observed and the syringe inserted more accurately through the open speculum. The plunger is then slowly pushed forward and the semen in the syringe is emptied deep into the vagina. The syringe (and speculum if used) may be left in place for several minutes and the woman is advised to lie still for about half-an-hour to improve the success rate.

Ordinary sexual lubricants should not be used in the process, but special fertility or 'sperm friendly' lubricants can be used for increased ease and comfort.

ICI is the easiest and most common insemination technique. The process closely replicates natural insemination, with fresh semen being directly deposited onto the neck of the cervix. It is the simplest artificial insemination method and unwashed or raw semen is normally used, but frozen semen (which has been thawed) may also be used. It is commonly used in home, self-insemination and practitioner insemination procedures, and for insemination where semen is provided by private donors. When performed at home without the presence of a professional this procedure is sometimes referred to as intravaginal insemination (IVI).

A conception cap, which is a form of conception device may be inserted into the vagina following insemination and may be left in place for several hours. Using this method, a woman may go about her usual activities while the cervical cap holds the semen in the vagina close to the entrance to the cervix. Advocates of this method claim that it increases the chances of conception. One advantage with the conception device is that fresh, non-liquefied semen may be used. The partner or donor may ejaculate straight into the cap and this can be immediately inserted into the vagina. Other methods may be used to insert semen into the vagina notably involving different uses of a conception cap. This may, for example, be inserted filled with sperm which does not have to be liquefied. The male may therefore ejaculate straight into the cap. Alternatively, a specially designed conception cap with a tube attached may be inserted empty into the vagina after which liquefied semen is poured into the tube. These methods are designed to ensure that donor or partner semen is inseminated as close as possible to the cervix and that it is kept in place there to increase the chances of conception.

Intrauterine Insemination

Intrauterine insemination (IUI) involves injection of washed sperm into the uterus with a catheter. If unwashed semen is used, it may elicit uterine cramping, expelling the semen and causing pain, due to content of prostaglandins. (Prostaglandins are also the compounds responsible for causing the myometrium to contract and expel the menses from the uterus, during menstruation.) Resting on the table for fifteen minutes after an IUI is optimal for the woman to increase the pregnancy rate.

Unlike ICI, intrauterine insemination normally requires a medical practitioner to perform the procedure. A female under 30 years of age has optimal chances with IUI; for the man, a TMS of more than 5 million per ml is optimal. In practice, donor sperm will satisfy these criteria. A promising cycle is one that offers two follicles measuring more than 16 mm, and estrogen of more than 500 pg/mL on the day of hCG administration. A short period of ejaculatory abstinence before intrauterine insemination is associated with higher pregnancy rates. However, GnRH agonist administration at the time of implantation does not improve pregnancy outcome in intrauterine insemination cycles according to a randomized controlled trial.

IUI is a more efficient method of artificial insemination. IUI can be used in conjunction with controlled ovarian hyperstimulation (COH). Still, advanced maternal age causes decreased success rates; women aged 38–39 years appear to have reasonable success during the first two cycles of ovarian hyperstimulation and IUI. However, for women aged over 40 years, there appears to be no benefit after a single cycle of COH/IUI. Medical experts therefore recommend considering in vitro fertilization after one failed COH/IUI cycle for women aged over 40 years.

A double intrauterine insemination theoretically increases pregnancy rates by decreasing the risk of missing the fertile window during ovulation. However, a randomized trial of insemination after ovarian hyperstimulation found no difference in live birth rate between single and double intrauterine insemination.

Intrauterine Tuboperitoneal Insemination

Intrauterine tuboperitoneal insemination (IUTPI) involves injection of washed sperm into both the uterus and fallopian tubes. The cervix is then clamped to prevent leakage to the vagina, best achieved with a specially designed double nut bivalve (DNB) speculum. The sperm is mixed to create a volume of 10 ml, sufficient to fill the uterine cavity, pass through the interstitial part of the tubes and the ampulla, finally reaching the peritoneal cavity and the Pouch of Douglas where it would be mixed with the peritoneal and follicular fluid. IUTPI can be useful in unexplained infertility, mild or moderate male infertility, and mild or moderate endometriosis. In non-tubal sub fertility, fallopian tube sperm perfusion may be the preferred technique over intrauterine insemination.

INTRATUBAL INSEMINATION

Intratubal insemination (ITI) involves injection of washed sperm into the fallopian tube, although this procedure is no longer generally regarded as having any beneficial effect compared with IUI. ITI however, should not be confused with gamete intrafallopian transfer, where both eggs and sperm are mixed outside the woman's body and then immediately inserted into the fallopian tube where fertilization takes place.

Pregnancy Rate

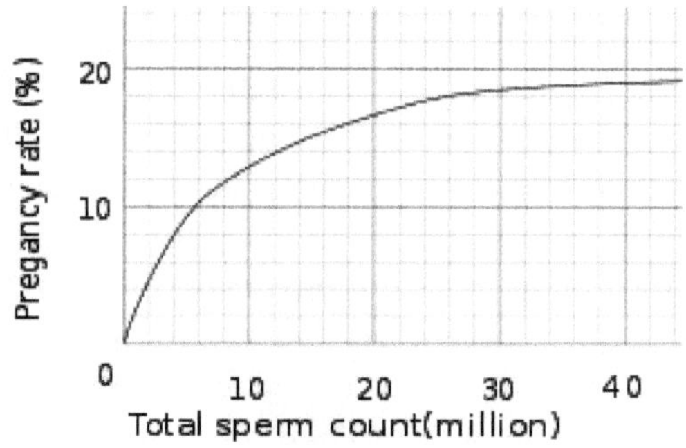

Approximate pregnancy rate as a function of total sperm count (may be twice as large as total motile sperm count). Values are for intrauterine insemination. (Old data, rates are likely higher today)

The pregnancy or success rates for artificial insemination are 10 to 15% per menstrual cycle using ICI, and 15–20% per cycle for IUI. In IUI, about 60 to 70% have achieved pregnancy after 6 cycles.

However, these pregnancy rates may be very misleading, since many factors, including the age and health of the recipient, have to be included to give a meaningful answer, e.g. definition of success and calculation of the total population. For couples with unexplained infertility, unstimulated IUI is no more effective than natural means of conception.

The pregnancy rate also depends on the total sperm count, or, more specifically, the total motile sperm count (TMSC), used in a cycle. The success rate increases with increasing TMSC, but only up to a certain count, when other factors become limiting to success. The summed pregnancy rate of two cycles using a TMSC of 5 million (may be a TSC of ~10 million on graph) in each cycle is substantially higher than one single cycle using a TMSC of 10 million. However, although more cost-efficient, using a lower TMSC also increases the average time taken to achieve pregnant. Women whose age is becoming a major factor in fertility may not want to spend that extra time.

Samples Per Child

The number of samples (ejaculates) required to give rise to a child varies substantially from person to person, as well as from clinic to clinic. However, the following equations generalize the main factors involved:

For **intracervical insemination**:

$$N = \frac{V_s \times c \times r_s}{n_r}$$

- N is how many children a single sample can give rise to.
- V_s is the volume of a sample (ejaculate), usually between 1.0 mL and 6.5 mL
- c is the concentration of motile sperm in a sample *after freezing and thawing*, approximately 5–20 million per ml but varies substantially
- r_s is the pregnancy rate per cycle, between 10% to 35%
- n_r is the total motile sperm count recommended for vaginal insemination (VI) or intra-cervical insemination (ICI), approximately 20 million pr. ml.

The pregnancy rate increases with increasing number of motile sperm used, but only up to a certain degree, when other factors become limiting instead.

Derivation of the Equation

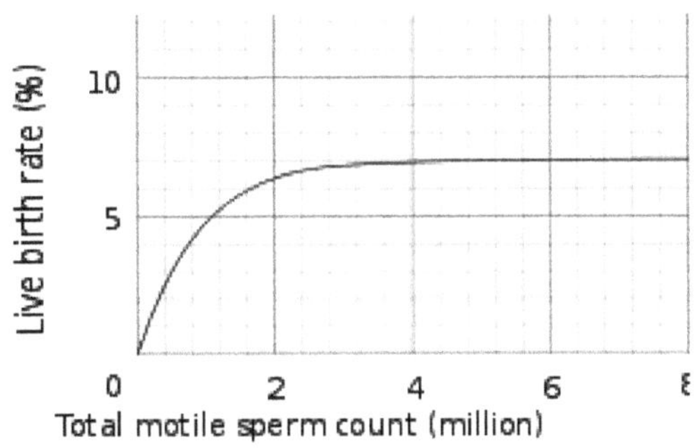

Approximate live birth rate (r_s) among infertile couples as a function of total motile sperm count (n_r). Values are for intrauterine insemination.

With these numbers, one sample would on average help giving rise to 0.1–0.6 children, that is, it actually takes on average 2–5 samples to make a child.

For **intrauterine insemination**, a *centrifugation fraction* (f_c) may be added to the equation:

f_c is the fraction of the volume that remains after centrifugation of the sample, which may be about half (0.5) to a third (0.33).

$$N = \frac{V_s \times f_c \times c \times r_s}{n_r}$$

On the other hand, only 5 million motile sperm may be needed per cycle with IUI (n_r=5 million)

Thus, only 1–3 samples may be needed for a child if used for IUI.

History

The first reported case of artificial insemination by donor occurred in 1884: a Philadelphia professor of medicine took sperm from his "best looking" student to inseminate an anesthetized woman. The woman was not informed about the procedure, unlike her infertile husband. The case was reported 25 years later in a medical journal. The sperm bank was developed in Iowa starting in the 1920s in research conducted by University of Iowa medical school researchers Jerome Sherman and Raymond Bunge.

In the 1980s, direct intraperitoneal insemination (DIPI) was occasionally used, where doctors injected sperm into the lower abdomen through a surgical hole or incision, with the intention of letting them find the oocyte at the ovary or after entering the genital tract through the ostium of the fallopian tube.

Social Implications

One of the key issues arising from the rise of dependency on assisted reproductive technology (ARTs) is the pressure placed on couples to conceive; 'where children are highly desired, parenthood is culturally mandatory, and childlessness socially unacceptable'.

The medicalization of infertility creates a framework in which individuals are encouraged to think of infertility quite negatively. In many cultures, especially those with large Muslim populations, donor insemination is religiously and culturally prohibited, often meaning that less accessible "high tech" and expensive ARTs, like IVF, are the only solution. An over-reliance on reproductive technologies in dealing with infertility prevents many – especially, for example, in the "infertility belt" of central and southern Africa – from dealing with many of the key causes of infertility treatable by artificial insemination techniques; namely preventable infections, dietary and lifestyle influences.

Morality in Christianity

An Anglican writer says that, "To achieve union but not children by means of contraceptives and to achieve children but not union by means of artificial insemination are both equally wrong." Heterosexual intercourse is viewed by the Catholic Church as a sacramental act meant to be experienced only by married couples; it is viewed as a physical representation of the spiritual unity of marriage between a husband and wife. According to the *Catechism of the Catholic Church*, artificial insemination "dissociates the sexual act from the procreative act. The act which brings the child into existence is no longer an act by two persons giving themselves to one another, but one that 'entrusts the life and identity of the embryo into the power of doctors and biologists and establishes the domination of technology over the origin and destiny of the human person. Such a relationship of domination is, in itself, contrary to the dignity and equality that must be common to parents and children'".

Legal Restrictions

Some countries restrict artificial insemination in a variety of ways. For example, some countries do not permit AI for single women, and some Muslim countries do not permit the use of donor sperm. As of May 2013, the following European countries permit medically assisted AI for single women:

- Armenia
- Belarus
- Belgium
- Bulgaria
- Cyprus
- Denmark
- Estonia
- Finland
- Germany
- Greece
- Hungary
- Iceland
- Ireland
- Latvia
- Macedonia
- Moldova
- Montenegro
- Netherlands
- Romania
- Russia
- Spain
- Ukraine
- United Kingdom

IN LIVESTOCK AND PETS

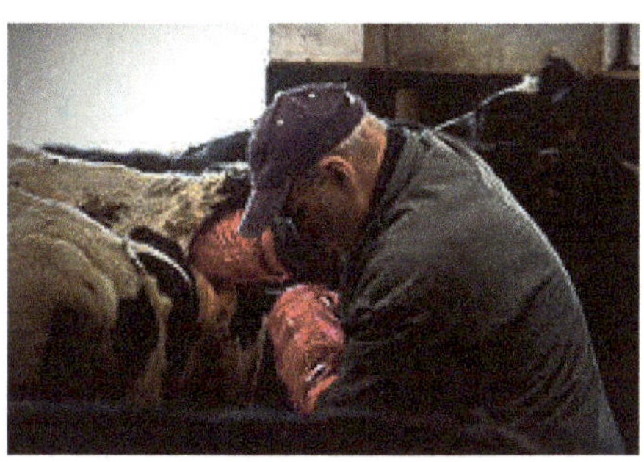

A man performing artificial insemination of a cow.

A breeding mount with built-in artificial vagina used in semen collection from horses for use in artificial insemination

The first viviparous to be artificially fertilized, was a dog. The experiment was conducted with success by the Italian Lazzaro Spallanzani in 1780. An important pioneer in artificial insemination was also the Russian Ilya Ivanov since 1899. In 1935 Suffolk sheep diluted semen was sent from Cambridge by plane to Krakoiv, Poland, and international research joint (Prawochenki from Poland, Milovanoff from URSS, Hammond from Cambridge, Walton from Scotland, and Thomasset from Uruguay).

Artificial insemination is used in many non-human animals, including sheep, horses, cattle, pigs, dogs, pedigree animals generally, zoo animals, turkeys and even honeybees. It may be used for many reasons, including to allow a male to inseminate a much larger number of females, to allow use of genetic material from males separated by distance or timc, to overcome physical breeding difficulties, to control the paternity of offspring, to synchronise births, to avoid injury incurred during natural mating, and to avoid the need to keep a male at all (such as for small numbers of females or in species whose fertile males may be difficult to manage).

Semen is collected, extended, then cooled or frozen. It can be used on site or shipped to the female›s location. If frozen, the small plastic tube holding the semen is referred to as a *straw*. To allow the sperm to remain viable during the time before and after it is frozen, the semen is mixed with a solution containing glycerol or other cryoprotectants. An *extender* is a solution that allows the semen from a donor to impregnate more females by making insemination possible with fewer sperm. Antibiotics, such as streptomycin, are sometimes added to the sperm to control some bacterial venereal diseases. Before the actual insemination, estrus may be induced through the use of progestogen and another hormone (usually PMSG or Prostaglandin F2α).

Artificial insemination of farm animals is very common in today›s agriculture industry in the developed world, especially for breeding dairy cattle (75% of all inseminations). Swine are also bred using this method (up to 85% of all inseminations). It provides an economical means for a livestock breeder to improve their herds utilizing males having very desirable traits.

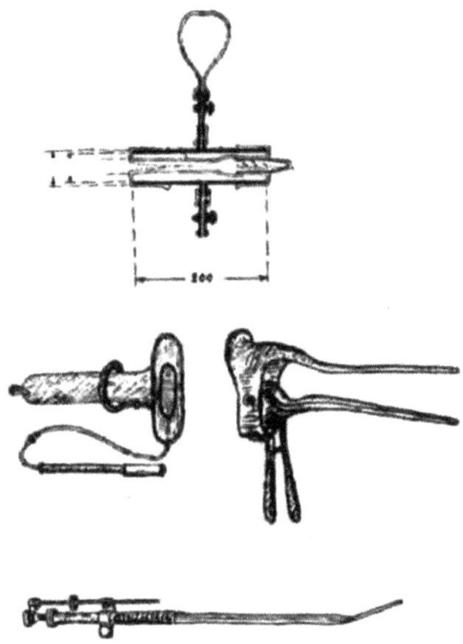

IA tools brought from the USSR by Luis Thomasset in 1935 to work at Cambridge Laboratories and South America.

Although common with cattle and swine, AI is not as widely practised in the breeding of horses. A small number of equine associations in North America accept only horses that have been conceived by «natural cover» or «natural service» – the actual physical mating of a mare to a stallion – the Jockey Club being the most notable of these, as no AI is allowed in Thoroughbred breeding. Other registries such as the AQHA and warmblood registries allow registration of foals created through AI, and the process is widely used allowing the breeding of mares to stallions not resident at the same facility – or even in the same country – through the use of transported frozen or cooled semen.

In modern species conservation, semen collection and artificial insemination is used also in birds. In 2013 scientist of the Justus-Liebig-University of Giessen, Germany, from the working group of Michael Lierz, Clinic for birds, reptiles, amphibians and fish, developed a novel technique for semen collection and artificial insemination in parrots producing the world›s first macaw by assisted reproduction (Lierz et al., 2013).

Modern artificial insemination was pioneered by John O. Almquist of the Pennsylvania State University. His improvement of breeding efficiency by the use of antibiotics (first proven with penicillin in 1946) to control bacterial growth, decreasing embrionic mortality and increase fertility, and various new techniques for processing, freezing and thawing of frozen semen significantly enhanced the practical utilization of AI in the livestock industry, and earned him the 1981 Wolf Foundation Prize in Agriculture. Many techniques developed by him have since been applied to other species, including that of the human male.

REFERENCES

1. European Sperm Bank USA
2. Robinson, Sarah (2010-06-24). "Professor". International Federation of Gynecology and Obstetrics. Retrieved 2012-12-27.
3. Adams, Robert, M.D."invitro fertilization technique", Monterey, CA, 1988
4. Laurie Barclay. "Immobilization May Improve Pregnancy Rate After Intrauterine Insemination". Medscape Medical News. Retrieved October 31, 2009.
5. Merviel P, Heraud MH, Grenier N, Lourdel E, Sanguinet P, Copin H (November 2008). "Predictive factors for pregnancy after intrauterine insemination (IUI): An analysis of 1038 cycles and a review of the literature". Fertil. Steril. 93 (1): 79–88. doi:10.1016/j.fertnstert.2008.09.058. PMID 18996517.
6. Marshburn PB, Alanis M, Matthews ML, et al. (September 2009). "A short period of ejaculatory abstinence before intrauterine insemination is associated with higher pregnancy rates". Fertil. Steril. 93 (1): 286–8. doi:10.1016/j.fertnstert.2009.07.972. PMID 19732887.
7. Bellver J, Labarta E, Bosch E, et al. (June 2009). "GnRH agonist administration at the time of implantation does not improve pregnancy outcome in intrauterine insemination cycles: a randomized controlled trial". Fertil. Steril. 94 (3): 1065–71. doi:10.1016/j.fertnstert.2009.04.044. PMID 19501354.
8. Harris, I.; Missmer, S.; Hornstein, M. (2010). "Poor success of gonadotropin-induced controlled ovarian hyperstimulation and intrauterine insemination for older women". Fertility and Sterility. 94 (1): 144–148. doi:10.1016/j.fertnstert.2009.02.040. PMID 19394605.
9. Bagis T, Haydardedeoglu B, Kilicdag EB, Cok T, Simsek E, Parlakgumus AH (May 2010). "Single versus double intrauterine insemination in multi-follicular ovarian hyperstimulation cycles: a randomized trial". Hum Reprod. 25 (7): 1684–90. doi:10.1093/humrep/deq112. PMID 20457669.
10. Leonidas Mamas, M.D. (March 2006). "Comparison of fallopian tube sperm perfusion and intrauterine tuboperitoneal insemination:a prospective randomized study". Fertility and Sterility. 85 (3): 735–740. doi:10.1016/j.fertnstert.2005.08.025. PMID 16500346.
11. G S Shekhawat, MD (2012). "Intrauterine insemination versus Fallopian tube sperm perfusion in non-tubal infertility". Internet Medical Journal.
12. Hurd WW, Randolph JF, Ansbacher R, Menge AC, Ohl DA, Brown AN (February 1993). "Comparison of intracervical, intrauterine, and intratubal techniques for donor insemination". Fertil. Steril. 59 (2): 339–42. PMID 8425628.
13. Utrecht CS News Subject: Infertility FAQ (part 4/4)
14. Intrauterine insemination. Information notes from the fertility clinic at Aarhus University Hospital, Skejby. By PhD Ulrik Kesmodel et al.
15. IVF.com
16. Fertility treatments 'no benefit'. BBC News, 7 August 2008

17. Bhattacharya S, Harrild K, Mollison J, et al. (2008). "Clomifene citrate or unstimulated intrauterine insemination compared with expectant management for unexplained infertility: pragmatic randomised controlled trial". BMJ. 337: a716. doi:10.1136/bmj.a716. PMC 2505091 . PMID 18687718.
18. Essig, Maria G. (2007-02-20). Van Houten, Susan; Landauer, Tracy, eds. "Semen Analysis". Healthwise. Reviewed by Martin Gabica and Avery L. Seifert. WebMD. Retrieved 2007-08-05.
19. Cryos International – What is the expected pregnancy rate (PR) using your donor semen? Archived January 7, 2011, at the Wayback Machine.
20. Cryos International – How much sperm should I order?
21. "Letter to the Editor: Artificial Impregnation". The Medical World: 163–164. April 1909. (cited in Gregoire, A.; Mayer, R. (1964). "The impregnators". Fertility and Sterility. 16: 130–134. PMID 14256095.)
22. Kara W. Swanson, "The Birth of the Sperm Bank," Annals of Iowa, 71 (Summer 2012), 241–76.
23. Oral Sex, a Knife Fight and Then Sperm Still Impregnated Girl. Account of a Girl Impregnated After Oral Sex Shows Sperms' Incredible Survivability By LAUREN COX. abcNEWS/Health Feb. 3, 2010
24. Cimino, C.; Guastella, G.; Comparetto, G.; Gullo, D.; Perino, A.; Benigno, M.; Barba, G.; Cittadini, E. (1988). "Direct intraperitoneal insemination (DIPI) for the treatment of refractory infertility unrelated to female organic pelvic disease". Acta Europaea fertilitatis. 19 (2): 61–68. PMID 3223194.
25. Marcia C. Inhorn (2003). "Global infertility and the globalization of new reproductive technologies: Illustrations from Egypt". Social Science and Medicine. 56: 1837–1851. doi:10.1016/s0277-9536(02)00208-3.
26. Joseph F. Fletcher (2015). Morals and Medicine: The Moral Problems of the Patient's Right to Know the Truth, Contraception, Artificial Insemination, Sterilization, Euthanasia. Princeton University Press. p. 101. ISBN 9781400868377.
27. "Paragraph 2377". Catechism of the Catholic Church, Second Edition. Libreria Editrice Vaticana. 2012.
28. Rainbow Europe Country Index[permanent dead link]
29. The Jockey Club has never allowed artificial insemination. ArchivedSeptember 20, 2008, at the Wayback Machine.
30. http://www.nature.com/srep/2013/130625/srep02066/full/srep02066.html
31. 1981 Wolf Foundation Prize in Agriculture

CHAPTER 5

Animal Sexual Behavior

Chapter Contents

Animal sexual Behavior takes many different forms, even within the same species. Researchers have observed monogamy, promiscuity, sex between species, sexual arousal from objects or places, sex apparently via duress or coercion, copulation with dead animals, homosexual, heterosexual and bisexual sexual Behavior, and situational sexual Behavior and a range of other practices among animals other than humans. Related studies have noted diversity in sexed bodies and gendered Behavior, such as intersex and transgender animals.

INTRODUCTION

Animal sexual Behavior takes many different forms, including within the same species. Common mating or reproductively motivated systems include monogamy, polyandry, polygamy, and promiscuity. Other sexual Behavior may be reproductively motivated (e.g. sex apparently due to duress or coercion and situational sexual Behavior) or non-reproductively motivated (e.g. interspecific sexuality, sexual arousal from objects or places, sex with dead animals, homosexual sexual Behavior, bisexual sexual Behavior).

When animal sexual Behavior is reproductively motivated, it is often termed *mating* or *copulation*; for most non-human mammals, mating and copulation occur at oestrus (the most fertile period in the mammalian female's reproductive cycle), which increases the chances of successful impregnation. Some animal sexual Behavior involves competition, sometimes fighting, between multiple males. Females often select males for mating only if they appear strong and able to protect themselves. The male that wins a fight may also have the chance to mate with a larger number of females and will therefore pass on his genes to their offspring.

Historically, it was believed that only humans and a small number of other species performed sexual acts other than for reproduction, and that animals› sexuality was instinctive and a simple "stimulus-response" Behavior. However, in addition to homosexual Behaviors, a range of species masturbate and may use objects as tools to help them do so. Sexual Behavior may be tied more strongly to establishment and maintenance of complex social bonds across a population which support its success in non-reproductive ways. Both reproductive and non-reproductive Behaviors can be related to expressions of dominance over another animal or survival within a stressful situation (such as sex due to duress or coercion).

Stags fighting – a common sexual Behavior

Greater sage-grouse at a lek, with multiple males displaying for the less conspicuous females.

Anatomical structures on the head and throat of a domestic turkey. 1. Caruncles, 2. Snood, 3. Wattle (dewlap), 4. Major caruncle, 5. Beard. During sexual Behavior, these structures enlarge or become brightly coloured.

MATING SYSTEMS

In sociobiology and Behavioral ecology, the term "mating system" is used to describe the ways in which animal societies are structured in relation to sexual Behavior. The mating system specifies which males mate with which females, and under what circumstances. There are four basic systems:

The four basic mating systems[4]:160–161[5]		
	Single female	**Multiple females**
Single male	Monogamy	Polygyny
Multiple males	Polyandry	Polygynandry

Monogamy

Monogamy occurs when one male mates with one female exclusively. A monogamous mating system is one in which individuals form long-lasting pairs and cooperate in raising offspring. These pairs may last for a lifetime, such as in pigeons, or it may occasionally change from one mating season to another, such as in emperor penguins. In contrast with tournament species, these pair-bonding species have lower levels of male aggression, competition and little sexual dimorphism. Zoologists and biologists now have evidence that monogamous pairs of animals are not always sexually exclusive. Many animals that form pairs to mate and raise offspring regularly engage in sexual activities with extra-pair partners. This includes previous examples, such as swans. Sometimes, these extra-pair sexual activities lead to offspring. Genetic tests frequently show that some of the offspring raised by a monogamous pair come from the female mating with an extra-pair male partner. These discoveries have led biologists to adopt new ways of talking about monogamy; According to Ulrich Reichard (2003):

Social monogamy refers to a male and female's social living arrangement (e.g., shared use of a territory, Behavior indicative of a social pair, and/or proximity between a male and female) without inferring any sexual interactions or reproductive patterns. In humans, social monogamy takes the form of monogamous marriage. Sexual monogamy is defined as an exclusive sexual relationship between a female and a male based on observations of sexual interactions. Finally, the term genetic monogamy is used when DNA analyses can confirm that a female-male pair reproduce exclusively with each other. A combination of terms indicates examples where levels of relationships coincide, e.g., sociosexual and sociogenetic monogamy describe corresponding social and sexual, and social and genetic monogamous relationships, respectively.

Whatever makes a pair of animals socially monogamous does not necessarily make them sexually or genetically monogamous. Social monogamy, sexual monogamy, and genetic monogamy can occur in different combinations.

Social monogamy is relatively rare in the animal kingdom. The actual incidence of social monogamy varies greatly across different branches of the evolutionary tree. Over 90% of avian species are socially monogamous. This stands in contrast to mammals. Only 3% of mammalian species are socially monogamous, although up to 15% of primate species are.Social monogamy has also been observed in reptiles, fish, and insects.

Sexual monogamy is also rare among animals. Many socially monogamous species engage in extra-pair copulations, making them sexually non-monogamous. For example, while over 90% of birds are socially monogamous, "on average, 30% or more of the baby birds in any nest are sired by someone other than the resident male." Patricia Adair Gowaty has estimated that, out of 180 different species of socially monogamous songbirds, only 10% are sexually monogamous.

The incidence of genetic monogamy, determined by DNA fingerprinting, varies widely across species. For a few rare species, the incidence of genetic monogamy is 100%, with all offspring genetically related to the socially monogamous pair. But genetic monogamy is strikingly low in other species. Barash and Lipton note:

The highest known frequency of extra-pair copulations are found among the fairy-wrens, lovely tropical creatures technically known as *Malurus splendens* and *Malurus cyaneus*. More than 65% of all fairy-wren chicks are fathered by males outside the supposed breeding group.

Such low levels of genetic monogamy have surprised biologists and zoologists, forcing them to rethink the role of social monogamy in evolution. They can no longer assume social monogamy determines how genes are distributed in a species. The lower the rates of genetic monogamy among socially monogamous pairs, the less of a role social monogamy plays in determining how genes are distributed among offspring.

Polygyny

Polygyny occurs when one male gets exclusive mating rights with multiple females. In some species, notably those with harem-like structures, only one of a few males in a group of females will mate. Technically, polygyny in sociobiology and zoology is defined as a system in which a male has a relationship with more than one female, but the females are predominantly bonded to a single male. Should the active male be driven out, killed, or otherwise removed from the group, in a number of species the new male will ensure that breeding resources are not wasted on another male's young. The new male may achieve this in many different ways, including:

- competitive infanticide: in lions, hippopotamuses, and some monkeys, the new male will kill the offspring of the previous alpha male to cause their mothers to become receptive to his sexual advances since they are no longer nursing.
- harassment to miscarriage: amongst wild horses and baboons, the male will "systematically harass" pregnant females until they miscarry.
- Pheromone-based spontaneous abortion
- in some rodents such as mice, a new male with a different scent will cause females who are pregnant to spontaneously fail to implant recently fertilised eggs. This does not require contact; it is mediated by scent alone. It is known as the Bruce effect.

Von Haartman specifically described the mating Behavior of the European pied flycatcher as successive polygyny. Within this system, the males leave their home territory once their primary female lays her first egg. Males then create a second territory, presumably in order to attract a secondary female to breed. Even when they succeed at acquiring a second mate, the males typically return to the first female to exclusively provide for her and her offspring.

Polygynous mating structures are estimated to occur in up to 90% of mammal species. As polygyny is the most common form of polygamy among vertebrates (including humans, to some extent), it has been studied far more extensively than polyandry or polygynandry.

Polyandry

The anglerfish *Haplophryne mollis* is polyandrous. This female is trailing the atrophied remains of males she has encountered.

Polyandry occurs when one female gets exclusive mating rights with multiple males. In some species, such as redlip blennies, both polygyny and polyandry are observed.

The males in some deep sea anglerfishes are much smaller than the females. When they find a female they bite into her skin, releasing an enzyme that digests the skin of their mouth and her body and fusing the pair down to the blood-vessel level. The male then slowly atrophies, losing first his digestive organs, then his brain, heart, and eyes, ending as nothing more than a pair of gonads, which release sperm in response to hormones in the female›s bloodstream indicating egg release. This extreme sexual dimorphism ensures that, when the female is ready to spawn, she has a mate immediately available. A single anglerfish female can "mate" with many males in this manner.

Polygynandry

Polygynandry occurs when multiple males mate indiscriminately with multiple females. The numbers of males and females need not be equal, and in vertebrate species studied so far, there are usually fewer males. Two examples of systems in primates are promiscuous mating chimpanzees and bonobos. These species live in social groups consisting of several males and several females. Each female copulates with many males, and vice versa. In bonobos, the amount of promiscuity is particularly striking because bonobos use sex to alleviate social conflict as well as to reproduce. This mutual promiscuity is the approach most commonly used by spawning animals, and is perhaps the "original fish mating system."[:161] Common examples are forage fish, such as herrings, which form huge mating shoals in shallow water. The water becomes milky with sperm and the bottom is draped with millions of fertilised eggs.

Polygamy

The term polygamy is an umbrella term used to refer generally to non-monogamous matings. As such, polygamous relationships can be polygynous, polyandrous or polygynandrous. In a small number of species, individuals can display either polygamous or monogamous Behavior depending on environmental conditions. An example is the social wasp *Apoica flavissima*. In some species, polygyny and polyandry is displayed by both sexes in the population. Polygamy in both sexes has been observed in red flour beetle (*Tribolium castaneum*).

A tournament species is one in which "mating tends to be highly polygamous and involves high levels of male-male aggression and competition." Tournament Behavior often correlates with high levels of sexual dimorphism, examples of species including chimpanzees and baboons. Most polygamous species present high levels of tournament Behavior, with a notable exception being bonobos.

Parental Investment and Reproductive Success

Mating grey slugs, suspended from a slime thread

Female and male sexual Behavior differ in many species. Often, males are more active in initiating mating, and bear the more conspicuous sexual ornamentation like antlers and colourful plumage. This is a result of anisogamy, where sperm are smaller and much less costly (energetically) to produce than eggs. This difference in physiological cost means that males are more limited by the number of mates they can secure, while females are limited by the quality of genes of her mates, a phenomenon known as Bateman's principle. Many females also have extra reproductive burdens in that parental care often falls mainly, or exclusively, on them. Thus, females are more limited in their potential reproductive success. In species where males take on more of the reproductive costs, such as sea horses and jacanas, the role is reversed, and the females are larger, more aggressive and more brightly coloured than the males.

In hermaphroditic animals, the costs of parental care can be evenly distributed between the sexes, e.g. earthworms. In some species of planarians, sexual Behavior takes the form of penis fencing. In this form of copulation, the individual that first penetrates the other with the penis, forces the other to be female, thus carrying the majority of the cost of reproduction. Post mating, banana slugs will sometimes gnaw off their partner's penis as an act of sperm competition. In the grey slug, the sharing of cost leads to a spectacular display, where the mates suspend themselves high above the ground from a slime thread, ensuring none of them can refrain from taking on the cost of egg-bearer.

Seasonality

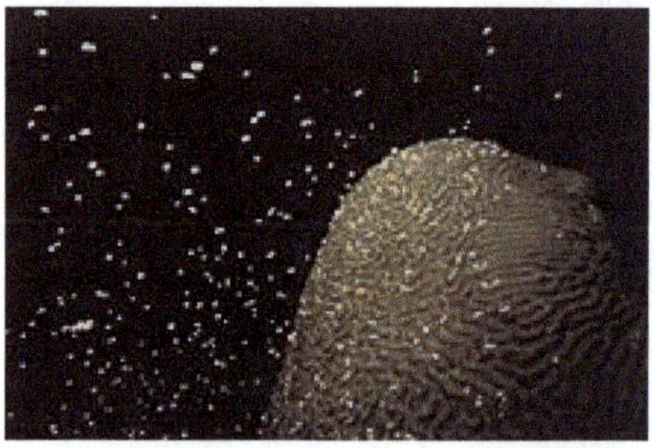

Brain corals typically spawning in connection with the full moon every August

Many animal species have specific mating (or breeding) periods e.g. (seasonal breeding) so that offspring are born or hatch at an optimal time. In marine species with limited mobility and external fertilisation like corals, sea urchins and clams, the timing of the common spawning is the only externally visible form of sexual Behavior. In areas with continuously high primary production, some species have a series of breeding seasons throughout the year. This is the case with most primates (who are primarily tropical and subtropical animals). Some animals (opportunistic breeders) breed dependent upon other conditions in their environment aside from time of year.

Mammals

Mating seasons are often associated with changes to herd or group structure, and Behavioral changes, including territorialism amongst individuals. These may be annual (e.g. wolves), biannual (e.g. dogs) or more frequently (e.g. horses). During these periods, females of most mammalian species are more mentally and physically receptive to sexual advances, a period scientifically described as estrous but commonly described as being "in season" or "in heat". Sexual Behavior may occur outside estrus, and such acts as do occur are not necessarily harmful.

Some mammals (e.g. domestic cats, rabbits and camilidae) are termed "induced ovulators". For these species, the female ovulates due to an external stimulus during, or just prior, to mating, rather than ovulating cyclically or spontaneously. Stimuli causing induced ovulation include the sexual Behavior of coitus, sperm and pheromones. Domestic cats have penile spines. Upon withdrawal of a cat's penis, the spines rake the walls of the female's vagina, which may cause ovulation.

Amphibians

For many amphibians, an annual breeding cycle applies, typically regulated by ambient temperature, precipitation, availability of surface water and food supply. This breeding season is accentuated in temperate regions, in boreal climate the breeding season is typically concentrated to a few short days in the spring.

Fish

Like many coral reef dwellers, the clownfish spawn around the time of the full moon in the wild. In a group of clownfish, there is a strict dominance hierarchy. The largest and most aggressive female is found at the top. Only two clownfish, a male and a female, in a group reproduce through external fertilisation. Clownfish are sequential hermaphrodites, meaning that they develop into males first, and when they mature, they become females. If the female clownfish is removed from the group, such as by death, one of the largest and most dominant males will become a female. The remaining males will move up a rank in the hierarchy.

MOTIVATION

Various neurohormones stimulate sexual wanting in animals. In general, studies have suggested that dopamine is involved in sexual incentive motivation, oxytocin and melanocortins in sexual attraction, and noradrenaline in sexual arousal. Vasopressin is also involved in the sexual Behavior of some animals.

Neurohormones in the Mating Systems of Voles

The mating system of prairie voles is monogamous; after mating, they form a lifelong bond. In contrast, montane voles have a polygamous mating style. When montane voles mate, they form no attachments and separate after copulation. Studies on the brains of these two species have found that it is two neurohormones and their respective receptors that are responsible for these differences in mating strategies. Male prairie voles release vasopressin after copulation prairie vole and attachment to the female then develops. Female prairie voles release oxytocin after reproducing and an attachment to her partner also develops. Neither male nor female montane voles have such a high quantity of oxytocin and vasopressin when they mate. Even when injected with these neurohormones, the mating system does not change. In contrast, if prairie voles are injected with the neurohormones, they may form a lifelong attachment, even if they have not mated. The reason for this is that prairie voles have more oxytocin and vasopressin receptors than do montane voles, and are therefore far more receptive to the two neurohormones. It is not the quantity of the hormone that determines the mating system and bond-formation, rather, it is number of neurohormone receptors.

Oxytocin and Rat Sexual Behavior

Mother rats experience a postparum estrus which makes them highly motivated to mate. However, they also have a strong motivation to protect their newly born pups. As a consequence, the mother rat solicits males to the nest but simultaneously becomes aggressive towards them to protect her young. If the mother rat is given injections of an oxytocin receptor antagonist, they no longer experience these maternal motivations.

PROLACTIN INFLUENCES SOCIAL BONDING IN RATS.

Oxytocin and Primate Sexual Behavior

Oxytocin plays a similar role in non-human primates as it does in humans.

Grooming, sex, and cuddling frequencies correlate positively with levels of oxytocin. As the level of oxytocin increases so does sexual motivation. While oxytocin plays a major role in parent child relationships, it is also found to play a role in adult sexual relationships. Its secretion affects the nature of the relationship or if there will even be a relationship at all.

Studies have shown that oxytocin is higher in monkeys in lifelong monogamous relationships compared to monkeys which are single. Furthermore, the oxytocin levels of the couples correlate positively; when the oxytocin secretion of one increases the other one also increases. Higher levels of oxytocin are related to monkeys expressing more Behaviors such as cuddling, grooming and sex, while lower levels of oxytocin reduce motivation for these activities.

Research on oxytocin's role in the animal brain suggests that it plays less of a role in Behaviors of love and affection than previously believed. "When oxytocin was first discovered in 1909, it was thought mostly to influence a mother's labour contractions and milk let-down. Then, in the 1990s, research with prairie voles found that giving them a dose of oxytocin resulted in the formation of a bond with their future mate (Azar, 40)." Oxytocin has since been treated by the media as the sole player in the "love and mating game" in mammals. This view, however, is proving to be false as, "most hormones don't influence Behavior directly. Rather, they affect thinking and emotions in variable ways (Azar, 40)." There is much more involved in sexual Behavior in the mammalian animal than oxytocin and vasopressin can explain.

Pleasure

It is often assumed that animals do not have sex for pleasure, or alternatively that humans, pigs (and perhaps dolphins and one or two species of primate) are the only species that do. This is sometimes stated as "animals mate only for reproduction". This view is considered a misconception by some scholars. Jonathan Balcombe argues that the prevalence of non-reproductive sexual Behavior in certain species suggests that sexual stimulation is pleasurable. He also points to the presence of the clitoris in some female mammals, and evidence for female orgasm in primates. On the other hand, it is impossible to know the subjective feelings of animals, and the notion that non-human animals experience emotions similar to humans is a contentious subject.

A 2006 Danish Animal Ethics Council report, which examined current knowledge of animal sexuality in the context of legal queries concerning sexual acts by humans, has the following comments, primarily related to domestically common animals:

Even though the evolution-related purpose of mating can be said to be reproduction, it is not actually the creating of offspring which originally causes them to mate. It is probable that they mate because they are motivated for the actual copulation, and because this is connected with a positive experience. It is therefore reasonable to assume that there is some form of pleasure or satisfaction connected with the act. This assumption is confirmed by the Behavior of males, who in the case of many species are prepared to work to get access to female animals, especially if the female animal is in oestrus, and males who for breeding purposes are used to having sperm collected become very eager, when the equipment they associate with the collection is taken out.

There is nothing in female mammals' anatomy or physiology that contradicts that stimulation of the sexual organs and mating is able to be a positive experience. For instance, the clitoris acts in the same way as with women, and scientific studies have shown that the success of reproduction is improved by stimulation of clitoris on (among other species) cows and mares in connection with insemination, because it improves the transportation of the sperm due to contractions of the inner genitalia. This probably also concerns female animals of other animal species, and contractions in the inner genitals are seen e.g. also during orgasm for women. It is therefore reasonable to assume that sexual intercourse may be linked with a positive experience for female animals.

KOINOPHILIA

Koinophilia is the love of the "normal" or phenotypically common (from the Greek, *koinos*, meaning "the usual" or "common"). The term was introduced to scientific literature in 1990, and refers to the tendency of animals seeking a mate to prefer that mate not to have any unusual, peculiar or deviant features. Similarly, animals preferentially choose mates with low fluctuating asymmetry. However, animal sexual ornaments can evolve through runaway selection, which is driven by (usually female) selection for non-standard traits.

Interpretation Bias

The field of study of sexuality in non-human species was a long-standing taboo. In the past, researchers sometimes failed to observe, mis-categorising and mis-described sexual Behavior which did not meet their preconceptions. In earlier periods, bias tended to support what would now be described as conservative sexual mores. An example of overlooking Behavior relates to descriptions of giraffe mating:

When nine out of ten pairings occur between males, "every male that sniffed a female was reported as sex, while anal intercourse with orgasm between males was only categorized as 'revolving around' dominance, competition or greetings."

In the 21st century, liberal social or sexual views are often projected upon animal subjects of research. Popular discussions of bonobos are a frequently cited example. Current research frequently expresses views such as that of the Natural History Museum at the University of Oslo, which in 2006 held an exhibition on animal sexuality:

Many researchers have described homosexuality as something altogether different from sex. They must realise that animals can have sex with who they will, when they will and without consideration to a researcher's ethical principles.

Other animal activities may be misinterpreted due to the frequency and context in which animals perform the Behavior. For example, domestic ruminants display Behaviors such as mounting and head-butting. This often occurs when the animals are establishing dominance relationships and are not necessarily sexually motivated. Careful analysis must be made to interpret what animal motivations are being expressed by those Behaviors.

TYPES OF SEXUAL BEHAVIOR

Reproductive Sexual Behavior

Cuckoldry

Small male bluegill sunfishes cuckold large males by adopting *sneaker* strategies.

Alternate male strategies which allow small males to engage in cuckoldry can develop in species such as fish where spawning is dominated by large and aggressive males. Cuckoldry is a variant of polyandry, and can occur with *sneak spawners*. A sneak spawner is a male that rushes in to join the spawning rush of a spawning pair. A spawning rush occurs when a fish makes a burst of speed, usually on a near vertical incline, releasing gametes at the apex, followed by a rapid return to the lake or sea floor or fish aggregation. Sneaking males do not take part in courtship. In salmon and trout, for example, *jack males* are common. These are small silvery males that migrate upstream along with the standard, large, hook-nosed males and that spawn by sneaking into redds to release sperm simultaneously with a mated pair. This Behavior is an evolutionarily stable strategy for reproduction, because it is favoured by natural selection just like the "standard" strategy of large males.

Hermaphroditism

Female groupers change their sex to male if no male is available.

Hermaphroditism occurs when a given individual in a species possesses both male and female reproductive organs, or can alternate between possessing first one, and then the other. Hermaphroditism is common in invertebrates but rare in vertebrates. It can be contrasted with gonochorism, where each individual in a species is either male or female, and remains that way throughout their lives. Most fish are gonochorists, but hermaphroditism is known to occur in 14 families of teleost fishes.

Usually hermaphrodites are *sequential*, meaning they can switch sex, usually from female to male (protogyny). This can happen if a dominant male is removed from a group of females. The largest female in the harem can switch sex over a few days and replace the dominant male. This is found amongst coral reef fishes such as groupers, parrotfishes and wrasses. It is less common for a male to switch to a female (protandry).[:162] As an example, most wrasses are protogynous hermaphrodites within a haremic mating system. Hermaphroditism allows for complex mating systems. Wrasses exhibit three different mating systems: polygynous, lek-like, and promiscuous mating systems.

Sexual Cannibalism

Araneus diadematus – cannibalistic mating Behavior

Sexual cannibalism is a Behavior in which a female animal kills and consumes the male before, during, or after copulation. Sexual cannibalism confers fitness advantages to both the male and female. Sexual cannibalism is common among insects, arachnids and amphipods. There is also evidence of sexual cannibalism in gastropods and copepods.

Sexual Coercion

During mating, the male muscovy duck typically immobilises the female.

Sex in a forceful or apparently coercive context has been documented in a variety of species. In some herbivorous herd species, or species where males and females are very different in size, the male dominates sexually by force and size.

Some species of birds appear to combine sexual intercourse with apparent violent assault; these include ducks, and geese.Female white-fronted bee-eaters are subjected to forced copulations. When females emerge from their nest burrows, males sometimes force them to the ground and mate with them. Such forced copulations are made preferentially on females who are laying and who may therefore lay eggs fertilised by the male.

In the *Acilius* genus of water beetles, an "evolutionary arms race" between the two sexes means that there is no courtship system for these beetles. Their mating Behaviors include males holding females underwater till exhausted, and allowing only occasional access to the surface to breathe for up to six hours (to prevent them breeding with other males), and females which have a variety of body shapings (to prevent males from gaining a grip).

It has been reported that young male elephants in South Africa sexually coerced and killed rhinoceroses. This interpretation of the elephants› Behavior was disputed by one of the original study›s authors, who said there was "nothing sexual about these attacks".

Parthenogenesis

Parthenogenesis is a form of asexual reproduction in which growth and development of embryos occur without fertilisation. Technically, parthenogenesis it is not a Behavior, however, sexual Behaviors may be involved.

Whip-tailed lizard females have the ability to reproduce through parthenogenesis and as such males are rare and sexual breeding non-standard. Females engage in "pseudocopulation"to stimulate ovulation, with their Behavior following their hormonal cycles; during low levels of oestrogen, these (female) lizards engage in "masculine" sexual roles. Those animals with currently high oestrogen levels assume "feminine" sexual roles. Lizards that perform the courtship ritual have greater fecundity than those kept in isolation due to an increase in hormones triggered by the sexual Behaviors. So, even though asexual whiptail lizards populations lack males, sexual stimuli still increase reproductive success. From an evolutionary standpoint these females are passing their full genetic code to all of their offspring rather than the 50% of genes that would be passed in sexual reproduction.

It is rare to find true parthenogenesis in fishes, where females produce female offspring with no input from males. All-female species include the Texas silverside, *Menidia clarkhubbsi* as well as a complex of Mexican mollies.:[162]

Parthenogenesis has been recorded in 70 vertebrate species including hammerhead sharks, blacktip sharks, amphibians and crayfish.

Unisexuality

Unisexuality occurs when a species is all-male or all-female. Unisexuality occurs in some fish species, and can take complex forms. *Squalius alburnoides*, a minnow found in several river basins in Portugal and Spain, appears to be an all-male species. The existence of this species illustrates the potential complexity of mating systems in fish. The species originated as a hybrid between two species, and is diploid, but not hermaphroditic. It can have triploid and tetraploid forms, including all-female forms that reproduce mainly through hybridogenesis.

Others

A dog mates with a coyote to produce a dog-coyote hybrid.

- **Interbreeding**: Hybrid offspring can result from the mating of two organisms of distinct but closely related parent species, although the resulting offspring is not always fertile. According to Alfred Kinsey, genetic studies on wild animal populations have shown a "large number" of inter-species hybrids.
- **Prostitution:** There are reports that animals occasionally engage in prostitution. A small number of pair-bonded females within a group of penguins took nesting material (stones) after copulating with a non-partner male. The researcher stated "I was watching opportunistically, so I can't give an exact figure of how common it really is." It has been reported that "bartering of meat for sex ... forms part of the social fabric of a troop of wild chimps living in the Tai National Park in the Cote d'Ivoire."
- **Pavlovian conditioning:** The sexualisation of objects or locations is recognised in the animal breeding world. For example, male animals may become sexually aroused upon visiting a location where they have been allowed to have sex before, or upon seeing a stimulus previously associated with sexual activity such as an artificial vagina. Sexual preferences for certain cues can be artificially induced in rats by pairing scents or objects with their early sexual experiences. The primary motivation of this Behavior is Pavlovian conditioning, and the association is due to a conditioned response (or association) formed with a distinctive "reward".
- **Viewing images:** A study using four adult male rhesus macaques (*Macaca mulatta*) showed that male rhesus macaques will give up a highly valued item, juice, to see images of the faces or perineum of high-status females. Encouraging captive pandas to mate is problematic. Showing young male pandas "panda pornography" is credited with a recent population boom among pandas in captivity in China. One researcher attributed the success to the sounds on the recordings.
- **Copulatory wounding and Traumatic Insemination**: Injury to a partner's genital tract during mating occurs in at least 40 taxa, ranging from fruit flies to humans. However, it often goes unnoticed due to its cryptic nature and because of internal wounds not visible outside.

Non-Reproductive Sexual Behavior

There is a range of Behaviors that animals perform which appear to be sexually motivated but which can not result in reproduction. These include:

- Masturbation: Some species, both male and female, masturbate, both when partners are available and otherwise.
- Oral sex: Several species engage in both autofellatio and oral sex. This has been documented in brown bears, Tibetan macaques, wolves, goats, primates, hyenas,bats, cape ground squirrels and sheep. In the greater short-nosed fruit bat, copulation by males is dorsoventral and the females lick the shaft or the base of the male›s penis, but not the glans which has already penetrated the vagina. While the females do this, the penis is not withdrawn and research has shown a positive relationship between length of the time that the penis is

licked and the duration of copulation. Post copulation genital grooming has also been observed.

- Homosexuality: Same-sex sexual Behavior occurs in a range of species, especially in social species, particularly in marine birds and mammals, monkeys, and the great apes. As of 1999, the scientific literature contained reports of homosexual Behavior in at least 471 wild species. Organisers of the Against Nature? exhibit stated that "homosexuality has been observed among 1,500 species, and that in 500 of those it is well documented."

A male black and white tegu mounts a female that has been dead for two days and attempts to mate.

- Genital-genital rubbing: This is sexual activity in which one animal rubs his or her genitals against the genitals of another animal. This is stated to be the "bonobo's most typical sexual pattern, undocumented in any other primate".
- Inter-species mating: Some animals opportunistically mate with individuals of another species.
- Sex involving juveniles: Male stoats (*Mustela erminea*) will sometimes mate with infant females of their species. This is a natural part of their reproductive biology – they have a delayed gestation period, so these females give birth the following year when they are fully grown. Juvenile male chimpanzees have been recorded mounting and copulating with immature chimps. Infants in bonobo societies are often involved in sexual Behavior.
- Necrophilia: This describes when an animal engages in a sexual act with a dead animal. It has been observed in mammals, birds, reptiles and frogs.
- Bisexuality: This describes when an animal shows sexual Behavior towards both males and females.
- Extended female sexuality: This is when females mate with males outside of their conceptive period.

Seahorse

Seahorses, once considered to be monogamous species with pairs mating for life, were described in a 2007 study as "promiscuous, flighty, and more than a little bit gay".

Scientists at 15 aquaria studied 90 seahorses of three species. Of 3,168 sexual encounters, 37% were same-sex acts. Flirting was common (up to 25 potential partners a day of both sexes); only one species (the British spiny seahorse) included faithful representatives, and for these 5 of 17 were faithful, 12 were not. Bisexual Behavior was widespread and considered "both a great surprise and a shock", with big-bellied seahorses of both sexes not showing partner preference. 1,986 contacts were male-female, 836 were female-female and 346 were male-male.

Bonobo

Bonobos mating, Jacksonville Zoo and Gardens.

The bonobo, which has a matriarchal society, is a fully bisexual species – both males and females engage in sexual Behavior with the same and the opposite sex, with females being particularly noted for engaging in sexual Behavior with each other and at up to 75% of sexual activity being nonreproductive. Primatologist Frans de Waal believes that bonobos use sexual activity to resolve conflict between individuals. Sexual activity occurs between almost all ages and sexes of bonobo societies.

Dolphin

Male bottlenose dolphins have been observed working in pairs to follow or restrict the movement of a female for weeks at a time, waiting for her to become sexually receptive. The same pairs have also been observed engaging in intense sexual play with each other. Janet Mann, a professor of biology and psychology at Georgetown University, argues that the common same-sex Behavior among male dolphin calves is about bond formation and benefits the species evolutionarily. They cite studies that have shown the dolphins later in life are bisexual and the male bonds forged from homosexuality work for protection as well as locating females with which to reproduce. In 1991, an English man was prosecuted for allegedly having sexual contact with a dolphin. The man was found not guilty after it was revealed at trial that the dolphin was known to tow bathers through the water by hooking his penis around them.

Hyena

The female spotted hyena has a unique urinary-genital system, closely resembling the penis of the male, called a pseudo-penis. The family structure is matriarchal and dominance relationships with strong sexual elements are routinely observed between related females. They are notable for using visible sexual arousal as a sign of submission but not dominance in males as well as females (females have a sizable erectile clitoris). It is speculated that to facilitate this, their sympathetic and parasympathetic nervous systems may be partially reversed in respect to their reproductive organs.

MATING BEHAVIOR

Vertebrates

Mammals

Mammals mate by vaginal copulation. To achieve this, the male usually mounts the female from behind. The female may exhibit lordosis in which she arches her back ventrally to facilitate entry of the penis. Amongst the land mammals, other than humans, only bonobos mate in a face-to-face position. Some sea mammals copulate in a belly-to-belly position. Some camelids mate in a lying-down position. In most mammals ejaculation occurs after multiple intromissions, but in most ruminant species, a single pelvic thrust occurs during copulation.

Invertebrates

Courting garden snails – the one on the left has fired a love dart into the one on the right.

A male star coral releases sperm into the water.

Invertebrates are often hermaphrodites. Some hermaphroditic land snails begin mating with an elaborate tactile courting ritual. The two snails circle around each other for up to six hours, touching with their tentacles, and biting lips and the area of the genital pore, which shows some preliminary signs of the eversion of the penis. As the snails approach mating, hydraulic pressure builds up in the blood sinus surrounding an organ housing a sharpened dart. The dart is made of calcium carbonate or chitin, and is called a love dart. Each snail manoeuvres to get its genital pore in the best position, close to the other snail's body. Then, when the body of one snail touches the other snail's genital pore, it triggers the firing of the love dart. After the snails have fired their darts, they copulate and exchange sperm as a separate part of the mating progression. The love darts are covered with a mucus that contains a hormone-like substance that facilitates the survival of the sperm.

Penis fencing is a mating Behavior engaged in by certain species of flatworm, such as *Pseudobiceros bedfordi*. Species which engage in the practice are hermaphroditic, possessing both eggs and sperm-producing testes. The species "fence" using two-headed dagger-like penises which are pointed, and white in colour. One organism inseminates the other. The sperm is absorbed through pores in the skin, causing fertilisation.

Corals can be both gonochoristic (unisexual) and hermaphroditic, each of which can reproduce sexually and asexually. Reproduction also allows corals to settle new areas. Corals predominantly reproduce sexually. 25% of hermatypic corals (stony corals) form single sex (gonochoristic) colonies, while the rest are hermaphroditic. About 75% of all hermatypic corals "broadcast spawn" by releasing gametes – eggs and sperm – into the water to spread offspring. The gametes fuse during fertilisation to form a microscopic larva called a planula, typically pink and elliptical in shape. Synchronous spawning is very typical on the coral reef and often, even when multiple species are present, all corals spawn on the same night. This synchrony is essential so that male and female gametes can meet. Corals must rely on environmental cues, varying from species to species, to determine the proper time to release gametes into the water. The cues involve lunar changes, sunset time, and possibly chemical signalling. Synchronous spawning may form hybrids and is perhaps involved in coral speciation.

Butterflies spend much time searching for mates. When the male spots a mate, he will fly closer and release pheromones. He then performs a special courtship dance to attract the female. If the female appreciates the dancing she may join him. Then they join their bodies together end to end at their abdomens. Here, the male passes the sperm to the female's egg-laying tube, which will soon be fertilised by the sperm. The male often dies shortly after mating.

Many animals make plugs of mucus to seal the female's orifice after mating. Normally such plugs are secreted by the male, to block subsequent partners. In spiders the female can assist the process. Spider sex is unusual in that males transfer their sperm to the female on small limbs called pedipalps. They use these to pick their sperm up from their genitals and insert it into the female's sexual orifice, rather than copulating directly. On the 14 occasions a sexual plug was made, the female produced it without assistance from the male. On ten of these occasions the male's pedipalps then seemed to get stuck while he was transferring the sperm (which is rarely the case in other species of spider), and he had great difficulty freeing himself. In two of those ten instances, he was eaten as a result.

GENETIC EVIDENCE OF INTERSPECIES SEXUAL ACTIVITY

Research into human evolution confirms that, in some cases, interspecies sexual activity may have been responsible for the evolution of new species (speciation). Analysis of animal genes found evidence that after humans had diverged from other apes, interspecies mating nonetheless occurred regularly enough to change certain genes in the new gene pool.Researchers found that the X chromosomes of humans and chimps may have diverged around 1.2 million years after the other chromosomes. One possible explanation is that modern humans emerged from a hybrid of human and chimp populations. A 2012 study questioned this explanation, concluding that "there is no strong reason to involve complicated factors in explaining the autosomal data".

Inbreeding Avoidance

When close relatives mate, progeny may exhibit the detrimental effects of inbreeding depression. Inbreeding depression is predominantly caused by the homozygous expression of recessive deleterious alleles. Over time, inbreeding depression may lead to the evolution of inbreeding avoidance Behavior. Several examples of animal Behavior that reduce mating of close relatives and inbreeding depression are described next.

Reproductively active female naked mole-rats tend to associate with unfamiliar males (usually non-kin), whereas reproductively inactive females do not discriminate. The preference of reproductively active females for unfamiliar males is interpreted as an adaptation for avoiding inbreeding.

When mice inbreed with close relatives in their natural habitat, there is a significant detrimental effect on progeny survival. In the house mouse, the major urinary protein (MUP) gene cluster provides a highly polymorphic scent signal of genetic identity that appears to underlie kin recognition and inbreeding avoidance. Thus there are fewer matings between mice sharing MUP haplotypes than would be expected if there were random mating.

Meerkat females appear to be able to discriminate the odour of their kin from the odour of their non-kin. Kin recognition is a useful ability that facilitates both cooperation among relatives and the avoidance of inbreeding. When mating does occur between meerkat relatives, it often results in inbreeding depression. Inbreeding depression was evident for a variety of traits: pup mass at emergence from the natal burrow, hind-foot length, growth until independence and juvenile survival.

The grey-sided vole (*Myodes rufocanus*) exhibits male-biased dispersal as a means of avoiding incestuous matings. Among those matings that do involve inbreeding the number of weaned juveniles in litters is significantly smaller than that from non-inbred litters indicating inbreeding depression.

In natural populations of the bird *Parus major* (great tit), inbreeding is likely avoided by dispersal of individuals from their birthplace, which reduces the chance of mating with a close relative.

Toads display breeding site fidelity, as do many amphibians. Individuals that return to natal ponds to breed will likely encounter siblings as potential mates. Although incest is possible, *Bufo americanus* siblings rarely mate. These toads likely recognise and actively avoid close kins as mates. Advertisement vocalisations by males appear to serve as cues by which females recognise their kin.

REFERENCES

1. Kent, Michael (2000). Advanced biology. Oxford University Press. pp. 250–253. ISBN 978-0-19-914195-1.
2. Thorpe, Showick; Thorpe, Edgar (2009). General Studies Manual. Pearson Education India. p. 17. ISBN 9788131721339.
3. Wickler, Wolfgang; Lorenz; Konrad; Kacher, Hermann (1974). "The sexual code : the social Behavior of animals and me".
4. Moyle PB and Cech JJ (2004) Fishes, An Introduction to Ichthyology. 5th Ed, Benjamin Cummings. ISBN 978-0-13-100847-2
5. Berglund A (1997) "Mating systems and sex allocation"Pages 237–265 in JJ Godon, ed. Behavioral ecology of teleost fishes. Oxford University Press. ISBN 0-19-850503-5.
6. "Pigeon". Encyclopædia Britannica. Encyclopædia Britannica. Retrieved 26 August 2014.
7. Lily Whiteman (13 February 2013). "Animal Attraction: The Many Forms of Monogamy in the Animal Kingdom". National Science Foundation. National Science Foundation. Retrieved 26 August 2014.
8. Ågren, G.; Zhou, Q.; Zhong, W. (1989). "Ecology and social Behavior of Mongolian gerbils, Meriones unguiculatus, at Xilinhot, Inner Mongolia, China". Animal Behavior. 37: 11–27. doi:10.1016/0003-3472(89)90002-X.
9. Birkhead, T.R.; Møller, A.P. (1995). "Extra-pair copulations and extra-pair paternity in birds". Animal Behavior. 49 (3): 843–848. doi:10.1016/0003-3472(95)80217-7.

10. Reichard, U.H. (2002). "Monogamy – a variable relationship" (PDF). Max Planck Research. 3: 62–67. Archived from the original (PDF) on 14 May 2011. Retrieved 24 April 2013.
11. Westneat, D. F.; Stewart, I. R. K. (2003). "EXTRA-PAIRPATERNITY INBIRDS: Causes, Correlates, and Conflict". Annual Review of Ecology, Evolution, and Systematics. 34: 365–396. doi:10.1146/annurev.ecolsys.34.011802.132439.
12. Birkhead, T.R. & Møller, A.P. (1996) "Monogamy and sperm competition in birds". In J. M. Black (Ed.), Partnerships in Birds: The Study of Monogamy, pp. 323–343, Oxford: Oxford University Press ISBN 0-19-854860-5.
13. Owens, I. P. F.; Hartley, I. R. (1998). "Sexual dimorphism in birds: Why are there so many different forms of dimorphism?". Proceedings of the Royal Society B: Biological Sciences. 265 (1394): 397–407. doi:10.1098/rspb.1998.0308. JSTOR 50849. PMC 1688905 .
14. Solomon, N. G.; Keane, B.; Knoch, L. R.; Hogan, P. J. (2004). "Multiple paternity in socially monogamous prairie voles (Microtus ochrogaster)". Canadian Journal of Zoology. 82 (10): 1667–1671. doi:10.1139/z04-142.
15. Reichard, U.H. (2003). Monogamy: Past and present. In U.H. Reichard and C. Boesch (Eds.), Monogamy: Mating strategies and partnerships in birds, humans, and other mammals, pp. 3–25, Cambridge: Cambridge University Press, ISBN 0521819733.
16. Barash, D.P. & Lipton, J.E. (2001). The Myth of Monogamy. New York, NY: W.H. Freeman and Company, ISBN 0805071369.
17. Angier, Natalie (21 August 1990). "Mating for Life? It›s Not for the Birds of the Bees". The New York Times.
18. Morell, V. (1998). "EVOLUTION OF SEX:A New Look at Monogamy". Science. 281 (5385): 1982–1983. doi:10.1126/science.281.5385.1982. PMID 9767050.
19. This section and examples taken from Robert Sapolsky (1998) Why Zebras Don›t Get Ulcers, W.H. Freeman and Co., ISBN 0-7167-3210-6, pp. 140–141.
20. Haartman, L. V. (1951). "Successive Polygamy". Behavior. 3: 256–273. doi:10.1163/156853951X00296.
21. Silverin, B. (1979). "Effects of long-acting testosterone administration on testes in free living pied flycatchers Ficedula hypoleuca". Endokrinologie. 74 (2): 141–146. PMID 583410.
22. Aloise King, Edith D.; Banks, Peter B.; Brooks, Robert C. (2011). "Sexual conflict in mammals: consequences for mating systems and life history". Mammal Review (published January 2013). 43 (1): 47–58. doi:10.1111/j.1365-2907.2011.00200.x.
23. Marraro, CH; Nursall JR (1983). "The reproductive periodicity and Behavior of Ophioblennius atlanticus at Barbados". J Zool. 61 (2): 317–325. doi:10.1139/z83-042.
24. Theodore W. Pietsch. "Precocious sexual parasitism in the deep sea ceratioid anglerfish, Cryptopsaras couesi Gill". Retrieved 31 July 2008.
25. Robert Sapolsky (2005). "Biology and Human Behavior: The Neurological Origins

of Individuality, 2nd edition". The Teaching Company. Retrieved 2010-11-10.

26. Bateman, A.J. (1948), "Intra-sexual selection in Drosophila", Heredity, 2 (Pt. 3): 349–368, doi:10.1038/hdy.1948.21, PMID 18103134
27. Trivers, R.L. (1972). Parental investment and sexual selection. In B. Campbell (Ed.), Sexual selection and the descent of man, 1871–1971 (pp. 136–179). Chicago, IL: Aldine. ISBN 0-435-62157-2.
28. Hermaphrodites duel for manhood, Science News Online. Accessed 14 March 2009.
29. Miller, Brooke L. W (2007). Sexual conflict and partner manipulation in the banana slug, Ariolimax dolichophallus. SANTA CRUZ: UNIVERSITY OF CALIFORNIA. Retrieved 13 February 2015.
30. Leonard, J. L. (5 May 2006). "Sexual selection: lessons from hermaphrodite mating systems". Integrative and Comparative Biology. 46 (4): 349–367. doi:10.1093/icb/icj041. PMID 21672747. Retrieved 11 February 2015.
31. For example, masturbation, trial mounting, and other Behaviors are regularly seen in male animals out of season
32. 2006 Danish Animal Ethics Council report: "The mucous membrane in the female animal's vagina and the animal's Behavior is under influence of its rut cycle. That means that the animal is physically and mentally more ready for sexual activities at some times than at others. But this does not mean that sexual activity will lead to injuries, fear or suffering, if it happens outside the rut period."
33. (Danish: "Slimhinden i hundyrets vagina og dyrets adfærd er under indflydelse af dets brunstcyklus. Det betyder, at dyret er fysisk og mentalt mere parat til seksuelle aktiviteter på nogle tidspunkter end på andre. Men dette er ikke ensbetydende med, at den seksuelle aktivitet vil være forbundet med skader, angst og lidelse, hvis den foregår udenfor brunstperioden.") Report November 2006 (PDF) Archived 12 March 2007 at the Wayback Machine.
34. Virginia Douglass Hayssen; Ari Van Tienhoven (1993). Asdell›s Patterns of Mammalian Reproduction: A Compendium of Species-specific Data. Cornell University Press. ISBN 978-0-8014-1753-5. Retrieved 27 September 2013.
35. Aronson, L. R.; Cooper, M. L. (1967). "Penile spines of the domestic cat: their endocrine-behavior relations." (PDF). Anatomical Record. 157 (1): 71–78. doi:10.1002/ar.1091570111. PMID 6030760.
36. Georgiadis, J. R.; Kringelbach, M. L. & Pfaus, J. G. (2012). "Sex for fun: a synthesis of human and animal neurobiology". Nature Reviews Urology. 9 (9): 486–498. doi:10.1038/nrurol.2012.151.
37. Bear, Mark F. (2007). Neuroscience, exploring the brain. Baltimore, MD: Lippincott Williams and Wilkins. pp. 544–545. ISBN 0781760038.
38. Kennett, J.E.; McKee, D.T. (2012). "Oxytocin: An emerging regulator of prolactin secretion in the female rat.". Journal of Neuroendocrinology. 24 (3): 403–412. doi:10.1111/j.1365-2826.2011.02263.x. PMC 3288386 . PMID 22129099.

39. Lambert, K. (2011). The Lab Rat Chronicles. New York: penguin group. pp. 151–172.

40. Azar, B. (March 2011). "Oxytocin›s other side". Science Watch. 42 (3): 40.

41. Aubert, Y.; Gustison, M. L.; Gardner, L. A.; Bohl, M. A.; Lange, J. R.; Allers, K. A.; Sommer, B.; Datson, N. A.; Abbott, D. H. (2012). "Flibanserin and 8-OH-DPAT Implicate Serotonin in Association between Female Marmoset Monkey Sexual Behavior and Changes in Pair-Bond Quality". The Journal of Sexual Medicine. 9 (3): 694–707. doi:10.1111/j.1743-6109.2011.02616.x. PMID 22304661.

42. Gil, M.; Bhatt, R.; Picotte, K. B.; Hull, E. M. (2011). "Oxytocin in the medial preoptic area facilitates male sexual behavior in the rat". Hormones and Behavior. 59 (4): 435–443. doi:10.1016/j.yhbeh.2010.12.012. PMC 3081415 . PMID 21195714.

43. Scott, Graham (2004). Essential Animal Behavior. Wiley-Blackwell. pp. 166–197. ISBN 0632057998.

44. Agrati, D.; Fernández-Guasti, A.; Ferreño, M.; Ferreira, A. (2011). "Coexpression of sexual behavior and maternal aggression: The ambivalence of sexually active mother rats toward male intruders". Behavioral Neuroscience. 125 (3): 446–451. doi:10.1037/a0023085. PMID 21517149.

45. McHenry, J. A.; Bell, G. A.; Parrish, B. P.; Hull, E. M. (2012). "Dopamine D1 receptors and phosphorylation of dopamine- and cyclic AMP-regulated phosphoprotein-32 in the medial preoptic area are involved in experience-induced enhancement of male sexual behavior in rats". Behavioral Neuroscience. 126 (4): 523–529. doi:10.1037/a0028707. PMC 3409344 . PMID 22708956.

46. Hull, Elaine M., R. L. Meisel, and B. D. Sachs. "Male sexual behavior." Hormones, brain and behavior 1 (2002): 3-137.

47. Jeffrey Moussaieff Masson (21 October 2009). When Elephants Weep: The Emotional Lives of Animals. Random House Publishing Group. ISBN 978-0-307-57420-6. Retrieved 28 May 2013.

48. Balcombe, J. (2009). "Animal pleasure and its moral significance". Applied Animal Behavior Science. 118 (3): 212. doi:10.1016/j.applanim.2009.02.012.

49. Dawkins, M. (2000). "Animal minds and animal emotions". American Zoologist. 40 (6): 883–888. doi:10.1668/0003-1569(2000)0400883:amaae2.0.co;2.

50. Panksepp, J. (1982). "Toward a general psychobiological theory of emotions". Behavioral and Brain Sciences. 5 (3): 407–422. doi:10.1017/S0140525X00012759.

51. "Emotions help animals to make choices (press release)". University of Bristol. 2010. Retrieved October 26, 2013.

52. Jacky Turner; Joyce D›Silva, eds. (2006). Animals, Ethics and Trade: The Challenge of Animal Sentience. Earthscan. ISBN 9781844072545. Retrieved October 26, 2013.

53. Det Dyreetiske Råd (2006). "Udtalelse om menneskers seksuelle omgang med dyr" (PDF). Justitsministeriet. Archived from the original (PDF) on 27 February

2012.

54. Det Dyreetiske Råd (2006). "Udtalelse om menneskers seksuelle omgang med dyr" (PDF). Justitsministeriet. Archived from the original (PDF) on 27 February 2012. Selv om det evolutionsmæssige formål med at parre sig kan siges at være reproduktion, er det ikke selve det, at dyrene får afkom, der i første omgang får dem til at parre sig. Det er til gengæld sandsynligt, at de parrer sig, fordi de er motiverede for selve parringsakten, og at denne er forbundet med en positiv oplevelse. Det er derfor rimeligt at antage, at der er en eller anden form for behag eller tilfredsstillelse forbundet med akten. Denne antagelse bekræftes af adfærden hos handyr, der for mange arters vedkommende er parate til at arbejde for at få adgang til hundyr, især hvis hundyret er i brunst, og handyr der i avlsøjemed er vant til at få tappet sæd – de viser stor ivrighed, når det udstyr, de forbinder med sædopsamlingen, tages frem.

55. Det Dyreetiske Råd (2006). "Udtalelse om menneskers seksuelle omgang med dyr" (PDF). Justitsministeriet. Archived from the original (PDF) on 27 February 2012. Der er intet ved hunpattedyrenes anatomi eller fysiologi, der modsiger, at stimulation af kønsorganerne og parring skulle kunne være en positiv oplevelse – fx fungerer klitoris på samme måde som hos kvinder. Videnskabelige undersøgelser har vist, at reproduktionssuccesen forbedres ved stimulation af klitoris på bl.a. køer og hopper i forbindelse med insemination, fordi det forbedrer sædtransporten pga. sammentrækninger af de indre kønsdele. Dette gælder sandsynligvis også hundyr af andre dyrearter, og sammentrækninger i de indre kønsdele ses fx også under orgasme hos kvinder. Det er derfor rimeligt at antage, at det seksuelle samvær kan være forbundet med en positiv oplevelse for hundyrene.

56. Koeslag, J.H. (1990). "Koinophilia groups sexual creatures into species, promotes stasis, and stabilizes social Behavior". Journal of Theoretical Biology. 144: 15–35. doi:10.1016/S0022-5193(05)80297-8. PMID 2200930.

57. Moller, A.P.; Pomiankowski, A (1993). "Fluctuating asymmetry and sexual selection". Genetica. 89: 267–279. doi:10.1007/bf02424520.

58. Dawkins, Richard (1986). The Blind Watchmaker. Longman, London. Published in Penguin Books 1988, 1991, and 2006. Chapter 8, Explosions and Spirals.

59. "1,500 animal species practice homosexuality". News-medical.net. 23 October 2006. Retrieved 19 February 2007.

60. Katz, L. S.; McDonald, T. J. (1992). "Sexual behavior of farm animals". Theriogenology. 38 (2): 239–253. doi:10.1016/0093-691X(92)90233-H. PMID 16727133.

61. Streak spawning Fishbase Glossary. Retrieved 11 February 2011.

62. Spawning rush Fishbase Glossary. Retrieved 11 February 2011.

63. Gross MR (1984) "Sunfish, salmon, and the evolution of alternative reproductive strategies and tactics in fishes". Pages 55–75 in GW Potts and RJ Wottoon, eds. Fish reproduction: Strategies and tactics. Academic Press.

64. Shapiro DY (1984) "Sex reversal and sociodemographics processes in coral reef fishes" Pages 103–116 in GW Potts and RK Wootoon, eds., Fish reproduction:

Strategies and tactics, Academic Press.

65. Robertson, D.R.; R.R. Warner. "Sexual patterns in the labroid fishes of the Western Caribbean II: the parrotfishes (Scaridae)". Smithsonian Contributions to Zoology. 255: 1–26. doi:10.5479/si.00810282.255.

66. Kazancioglu, E.; S.H. Alonzo (August 2010). "A comparative analysis of sex change in Labridae supports the size advantage hypothesis". Evolution. 64 (8): 2254–2264. doi:10.1111/j.1558-5646.2010.01016.x. PMID 20394662.

67. Colin, P.L.; L. J. Bell. "Aspects of the spawning of labrid and scarid fishes (Pisces, Labroidei) at Enewetak Atoll, Marshall Islands with notes on other families (corrected reprint.)". Environmental Biology of Fishes. 33 (3): 330–345. doi:10.1007/BF00005881.

68. Polis, G.A. (1981). "The evolution and dynamics of intraspecific +4193 predation". Annual Reviews Ecological Systems. 51: 225–251.

69. Bilde, T.; Tuni, C.; Elsayed, R.; Pekár, S.; Toft, S. (2006). "Death feigning in the face of sexual cannibalism". Biology Letters. 2: 23–5. doi:10.1098/rsbl.2005.0392.

70. Bailey, R.O.; Seymour, N. R.; Stewart, G.R. (1978). "Rape Behavior in blue-winged teal" (PDF). Auk. 95: 188–90. doi:10.2307/4085514. JSTOR 4085514.

71. Barash, D. P. (1977). "Sociobiology of Rape in Mallards (Anas platyrhynchos): Responses of the Mated Male". Science. 197 (4305): 788–789. Bibcode:1977Sci...197..788B. doi:10.1126/science.197.4305.788. PMID 17790773.

72. Emlen, S. T.; Wrege, P. H. (2010). "Forced Copulations and Intra-specific Parasitism: Two Costs of Social Living in the White-fronted Bee-eater". Ethology. 71: 2–29. doi:10.1111/j.1439-0310.1986.tb00566.x.

73. "Not tonight, not ever. I've got a headache. Don't come near me", The Times, 25 June 2007 p. 25

74. Siebert, Charles. (2006-10-08) ‹›An Elephant Crackup?››, Charles Siebert, New York Times Magazine, October 8, 2006. Nytimes.com. Retrieved on 2011-12-22.

75. "Have elephants begun raping rhinos in the wild?". The Straight Dope. 25 July 2008. Retrieved 5 March 2013.

76. Hiskey, D. "New Mexico whiptail lizards are all females". Retrieved February 13, 2015.

77. Echelle AA, Echelle AF, Crozier CD (1983). "Evolution of an all-female fish, Menidia clarkhubbsi (Atherinidae)". Evolution. 37 (4): 772–784. doi:10.2307/2407918.

78. Harmon, K. (2010). "No sex needed: All-female lizard species cross their chromosomes to make babies.". Scientific American. Retrieved February 13, 2015.

79. "Captive shark had ‹virgin birth›". BBC News. 2007-05-23. Retrieved 23 December 2008.

80. "›Virgin birth› for aquarium shark". Metro.co.uk. 2008-10-10. Retrieved 2008-10-10.

81. Halliday, Tim R.; Adler, Kraig, eds. (1986). Reptiles & Amphibians. Torstar Books. p. 101. ISBN 0-920269-81-8.

82. Walker, Brian (2010-11-11). "Scientists discover unknown lizard species at lunch buffet". CNN. Retrieved 2010-11-11.

83. Scholtz, Gerhard; Braband, Anke; Tolley, Laura; Reimann, André; Mittmann, Beate; Lukhaup, Chris; Steuerwald, Frank; Vogt, GüNter (2003). "Ecology: Parthenogenesis in an outsider crayfish". Nature. 421 (6925): 806. Bibcode:2003Natur.421..806S. doi:10.1038/421806a. PMID 12594502.

84. Martin, Peer; Kohlmann, Klaus; Scholtz, Gerhard (2007). "The parthenogenetic Marmorkrebs (marbled crayfish) produces genetically uniform offspring". Naturwissenschaften. 94 (10): 843–6. Bibcode:2007NW.....94..843M. doi:10.1007/s00114-007-0260-0. PMID 17541537.

85. Alves MJ, Collarea-Pereira MJ, Dowling TE, Coelho MM (2002). "The genetics of maintenance of an all-male lineage in the Squalius alburnoides complex". J. Fish. Biol. 60: 649–662. doi:10.1111/j.1095-8649.2002.tb01691.x.

86. Kinsey, Alfred (1953). Sexual Behavior in the Human Female. W.B. Saunders Company. p. 504.

87. "Penguins are turning to prostitution". BBC. 26 February 1998. Retrieved 12 June 2008.

88. Connor, Steve (8 April 2009). "Sex for meat – how chimps seduce their mates". The Independent. London.

89. Pfaus, J. G.; Kippin, T. E.; Coria-Avila, G. A.; Gelez, H.; Afonso, V. M.; Ismail, N.; Parada, M. (2012). "Who, what, where, when (and maybe even why)? How the experience of sexual reward connects sexual desire, preference, and performance". Archives of Sexual Behavior. 41 (1): 31–62. doi:10.1007/s10508-012-9935-5.

90. Deaner M. O.; Khera A. V.; Platt M. L. (2005). "Monkeys pay per view: adaptive valuation of social images by rhesus macaques" (PDF). Current Biology. 15 (6): 543–548. doi:10.1016/j.cub.2005.01.044. PMID 15797023.

91. Gray, Denis D. (27 November 2006) Porn sparks panda baby boom in China: Research — and blue movies — attributed to record-high birth rate in 2006. Associated Press (via MSNBC). Retrieved on 22 December 2011.

92. Klaus Reinhardt, Nils Anthes, and Rolanda Lange (2015) Copulatory Wounding and Traumatic Insemination" Cold Spring Harb Perspect Biol 2015; 7: a017582

93. Watson, P. F. (1978). Artificial breeding of non-domestic animals: (the proceedings of a symposium held at the Zoological Society of London on 7 and 8 September 1977). Academic Press for the Zoological Society of London. ISBN 978-0-12-613343-1. Retrieved 9 February 2013.

94. Balcombe, Jonathan P. (2011). The Exultant Ark: A Pictorial Tour of Animal Pleasure. University of California Press. pp. 89–. ISBN 978-0-520-26024-5.

95. These Bears Are Having Lots Of Oral Sex, And Scientists Think They Know Why (The Huffington Post) By: Grenoble, Ryan.

96. Hideshi Ogawa (2006). Wily Monkeys: Social Intelligence of Tibetan Macaques. Kyoto University Press. pp. 4–. ISBN 978-1-920901-97-4.

97. Fox, M. W. (1972). "The Social Significance of Genital Licking in the Wolf, Canis lupus". Journal of Mammalogy. 53: 637–640. doi:10.2307/1379064. JSTOR 1379064.

98. G. R. Pafumi (15 January 2010). Is Our Vision of God Obsolete?. Xlibris Corporation. ISBN 978-1-4500-0396-4.

99. Tan, M.; Jones, G.; Zhu, G.; Ye, J.; Hong, T.; Zhou, S.; Zhang, S.; Zhang, L. (2009). Hosken, David, ed. "Fellatio by Fruit Bats Prolongs Copulation Time". PLoS ONE. 4 (10): e7595. Bibcode:2009PLoSO...4.7595T. doi:10.1371/journal.pone.0007595. PMC 2762080 . PMID 19862320.

100. Waterman, J. M. (2010). Briffa, Mark, ed. "The Adaptive Function of Masturbation in a Promiscuous African Ground Squirrel". PLoS ONE. 5 (9): e13060. Bibcode:2010PLoSO...513060W. doi:10.1371/journal.pone.0013060. PMC 2946931 . PMID 20927404.

101. Tan, Min; Gareth Jones; Guangjian Zhu; Jianping Ye; Tiyu Hong; Shanyi Zhou; Shuyi Zhang; Libiao Zhang (28 October 2009). Hosken, David, ed. "Fellatio by Fruit Bats Prolongs Copulation Time". PLoS ONE. 4 (10): e7595. Bibcode:2009PLoSO...4.7595T. doi:10.1371/journal.pone.0007595. PMC 2762080 . PMID 19862320. Retrieved 28 October 2009.

102. Bagemihl, Bruce (1999). Biological Exuberance: Animal Homosexuality and Natural Diversity. St. Martin›s Press. p. 673.

103. "Oslo gay animal show draws crowds". BBC News. 19 October 2006. Archived from the original on 29 October 2006. Retrieved 19 October 2006.

104. Sazima, I. (2015). "Corpse bride irresistible: a dead female tegu lizard (Salvator merianae) courted by males for two days at an urban park in South-eastern Brazil". Herpetology Notes. 8: 15–18.

105. de Waal FB (1995). "Bonobo sex and society". Sci Am.272 (3): 82–8. doi:10.1038/scientificamerican0395-82. PMID 7871411. Perhaps the bonobo›s most typical sexual pattern, undocumented in any other primate, is genito-genital rubbing (or GG rubbing) between adult females. One female facing another clings with arms and legs to a partner that, standing on both hands and feet, lifts her off the ground

106. Paoli, T.; Palagi, E.; Tacconi, G.; Tarli, S. B. (2006). "Perineal swelling, intermenstrual cycle, and female sexual behavior in bonobos (Pan paniscus)". American Journal of Primatology. 68 (4): 333–347. doi:10.1002/ajp.20228. PMID 16534808.

107. Walker, Matt (2 May 2008) Science/Nature ‹Sex pest› seal attacks penguin. BBC News. Retrieved on 15 February 2011.

108. Doncarlos, Michael W.; Petersen, Jay S.; Tilson, Ronald L. (1986). "Captive biology of an asocial mustelid; Mustela erminea". Zoo Biology. 5 (4): 363–370. doi:10.1002/zoo.1430050407.

109. Dawkins, Richard (2004). "Chimpanzees". The Ancestor›s Tale. Houghton Mifflin. ISBN 1-155-16265-X.

110. de Mattos Brito, L. B.; Joventino, I. R.; Ribeiro, S. C.; Cascon, P. (2012). "Necrophiliac behavior in the "cururu" toad,Rhinella jimi Steuvax, 2002, (Anura, Bufonidae) from Northeastern Brazil" (PDF). North-Western Journal of Zoology. 8 (2): 365.

111. Thornhill, R.; Gangestad, S. W. (2008). The Evolutionary Biology of Human Female Sexuality. USA: Oxford University Press. pp. 37–55.

112. Simon De Bruxelles Promiscuous and bisexual — the ‹faithful› seahorse has a secret sex life. Timesonline.co.uk. 31 January 2007. Retrieved on 22 December 2011.

113. Homosexual Activity Among Animals Stirs Debate, National Geographic (22 July 2004) Retrieved 6 November 2007

114. Zihlman, A. L.; Hunter, W. S. (1972). "A biomechanical interpretation of the pelvis of Australopithecus". Folia Primatologica. 18 (1): 1–19. doi:10.1159/000155465. PMID 4658666.

115. Central Park Zoo›s gay penguins ignite debate. Sfgate.com (7 February 2004). Retrieved on 22 December 2011.

116. Unwin, Brian (22 January 2008). "›Tougher laws› to protect friendly dolphins". London: The Telegraph.

117. East, Marion L.; Hofer, Heribert; Wickler, Wolfgang (1993). "The erect ‹penis› is a flag of submission in a female-dominated society: greetings in Serengeti spotted hyenas". Behavioral Ecology and Sociobiology. 33 (6): 355–370. doi:10.1007/BF00170251.

118. Sapolsky (1998), Why Zebras Don›t Get Ulcers, W.H. Freeman and Co., ISBN 0-7167-3210-6, pp. 127–129.

119. Dewsbury, Donald A. "Patterns of copulatory behavior in male mammals." Quarterly Review of Biology (1972): 1-33.

120. Fortey, I. (2008). "The 15 most bizarre animal mating rituals". Retrieved February 18, 2015.

121. Diaz, K. "Tursiops aduncus Indo-Pacific bottlenose dolphin". Animal Diversity Web. Retrieved October 19, 2016.

122. Adulyanukosol, K.; Thongsukdee, S.; Hara, T.; Arai, N.; Tsuchiya, M. (2007). "Observations of dugong reproductive behavior in Trang Province, Thailand: further evidence of intraspecific variation in dugong behavior". Marine Biology. 151 (5): 1887–1891. doi:10.1007/s00227-007-0619-y.

123. "Bactrian & Dromedary Camels". Factsheets. San Diego Zoo Global Library. March 2009. Retrieved 4 December 2012.

124. Encyclopedia of Behavioral Neuroscience. Elsevier Science. 16 April 2010. ISBN 978-0-08-091455-8.

125. David E. Noakes (23 April 2009). Arthur›s Veterinary Reproduction and Obstetrics. Elsevier Health Sciences UK. pp. 714–. ISBN 978-0-7020-3990-4.

126. Donald M. Broom; Andrew Ferguson Fraser (1 January 2007). Domestic Animal Behavior and Welfare. CABI. pp. 156–. ISBN 978-1-84593-287-9.

127. Rogers, David; Chase, Ronald (2001). "Dart receipt promotes sperm storage in the garden snail Helix aspersa". Behavioral Ecology and Sociobiology. 50 (2): 122–7. doi:10.1007/s002650100345.

128. Chase, R. (2007). "The function of dart shooting in helicid snails". American Malacological Bulletin. 23: 183–189. doi:10.4003/0740-2783-23.1.183.

129. Chase, R.; Blanchard, K. C. (2006). "The snail›s love-dart delivers mucus to increase paternity". Proceedings of the Royal Society of London B, Biological Sciences. 273 (1593): 1471–1475. doi:10.1098/rspb.2006.3474.

130. Newman, Leslie. "Fighting to mate: flatworm penis fencing". PBS.

131. Veron, J.E.N. (2000). Corals of the World. Vol 3 (3rd ed.). Australia: Australian Institute of Marine Sciences and CRR Qld Pty Ltd. ISBN 0-642-32236-8.

132. Barnes, R. and; Hughes, R. (1999). An Introduction to Marine Ecology (3rd ed.). Malden, MA: Blackwell Science, Inc. pp. 117–141. ISBN 0-86542-834-4.

133. Hatta, M.; Fukami, H.; Wang, W.; Omori, M.; Shimoike, K.; Hayashibara, T.; Ina, Y.; Sugiyama, T. (1999). "Reproductive and genetic evidence for a reticulate evolutionary theory of mass spawning corals" (PDF). Molecular Biology and Evolution. 16 (11): 1607–1613. doi:10.1093/oxfordjournals.molbev.a026073. PMID 10555292.

134. "Chan Lee Peng, August 2008". Scienceray.com. Retrieved 5 March 2013.

135. Sexual Appetite and Animal Behavior – Sex and the single spider. 400. The Economist Intelligence Unit. 14 July 2011. ISSN 0013-0613.

136. Patterson, Nick; Daniel J. Richter; Sante Gnerre; Eric S. Lander; David Reich (June 29, 2006). "Genetic evidence for complex speciation of humans and chimpanzees". Nature. 441 (7097): 1103–1108. Bibcode:2006Natur.441.1103P. doi:10.1038/nature04789. ISSN 0028-0836. PMID 16710306.

137. Wade, Nicholas (18 May 2006) Two Splits Between Human and Chimp Lines Suggested, New York Times.

138. Yamamichi, M; Gojobori J; Innan H. (Jan 2012). "An autosomal analysis gives no genetic evidence for complex speciation of humans and chimpanzees". Mol Biol Evol. 29 (1): 145–56. doi:10.1093/molbev/msr172. PMID 21903679.

139. Charlesworth D, Willis JH (2009). "The genetics of inbreeding depression". Nat. Rev. Genet. 10 (11): 783–96. doi:10.1038/nrg2664. PMID 19834483.

140. Clarke FM, Faulkes CG (1999). "Kin discrimination and female mate choice in the naked mole-rat Heterocephalus glaber". Proc. Biol. Sci. 266 (1432): 1995–2002. doi:10.1098/rspb.1999.0877. PMC 1690316 . PMID 10584337.

141. Jiménez JA, Hughes KA, Alaks G, Graham L, Lacy RC (1994). "An experimental study of inbreeding depression in a natural habitat". Science. 266 (5183): 271–3. doi:10.1126/science.7939661. PMID 7939661.

142. Sherborne AL, Thom MD, Paterson S, Jury F, Ollier WE, Stockley P, Beynon RJ, Hurst JL (2007). "The genetic basis of inbreeding avoidance in house mice".

Curr. Biol. 17 (23): 2061–6. doi:10.1016/j.cub.2007.10.041. PMC 2148465 . PMID 17997307.

143. Leclaire S, Nielsen JF, Thavarajah NK, Manser M, Clutton-Brock TH (2013). "Odour-based kin discrimination in the cooperatively breeding meerkat". Biol. Lett. 9 (1): 20121054. doi:10.1098/rsbl.2012.1054. PMC 3565530 . PMID 23234867.

144. Nielsen JF, English S, Goodall-Copestake WP, Wang J, Walling CA, Bateman AW, Flower TP, Sutcliffe RL, Samson J, Thavarajah NK, Kruuk LE, Clutton-Brock TH, Pemberton JM (2012). "Inbreeding and inbreeding depression of early life traits in a cooperative mammal". Mol. Ecol. 21 (11): 2788–804. doi:10.1111/j.1365-294X.2012.05565.x. PMID 22497583.

145. Ishibashi Y, Saitoh T (2008). "Role of male-biased dispersal in inbreeding avoidance in the grey-sided vole (Myodes rufocanus)". Mol. Ecol. 17 (22): 4887–96. doi:10.1111/j.1365-294X.2008.03969.x. PMID 19140979.

146. Szulkin M, Sheldon BC (2008). "Dispersal as a means of inbreeding avoidance in a wild bird population". Proc. Biol. Sci. 275 (1635): 703–11. doi:10.1098/rspb.2007.0989. PMC 2596843 . PMID 18211876.

147. Waldman, B; Rice, JE; Honeycutt, RL (1992). "Kin recognition and incest avoidance in toads". Am Zool. 32: 18–30. doi:10.1093/icb/32.1.18.

CHAPTER 6

Breed Standards

Chapter Contents

- **Introduction**
- **Contents of a Breed Standard**
- **Examples of Use**

A breed standard is a list of points used to describe a breed of animal, such as a cat, dog, cow, horse, goat, or sheep, among many others. The guidelines in a breed standard are used to determine which animals will be accepted or admitted into the breed, and which ones will be rejected. Breed standards are designed to protect the integrity of purebred animals and their heritage by ensuring that only the finest representatives of the breed are identified as members of the breed.

INTRODUCTION

A breed standard (also called bench standard or standard of points) in animal fancy and animal husbandry is a set of guidelines which is used to ensure that the animals produced by a breeder or breeding facility conform to the specifics of the standardized breed.

Breed standards are devised by breed associations or breed clubs, not by individuals, and are written to reflect the use or purpose of the species and breed of the animal. Breed standards help define the ideal animal of a breed and provide goals for breeders in improving stock. In essence a breed standard is a blueprint for an animal fit for the function it was bred - i.e. herding, tracking etc. Breed standards are not scientific documents, and may vary from association to association, and from country to country, even for the same species and breed. There is no one format for breed standards across all species, and breed standards do change and are updated over time.

CONTENTS OF A BREED STANDARD

Breed standards cover the *externally observable* qualities of the animal such as *appearance*, *movement*, and *temperament*. The exact format of the breed standard varies, as breed standards are not scientific documents and change as the needs of the members of the organization which authors them change. In general, a breed standard may include history of the breed, a narrative description of the breed, and details of the ideal externally observable structure and behavior for the breed. Certain deviations from the standard are considered *faults*. A large degree of deviation from the breed standard, an excess of faults, or certain defined major faults, may indicate that the animal should not be bred, although its fitness for other uses may not be impeded by the faults. An animal that closely matches (conforms to) the breed standard for its species and breed is said to have good conformation.

EXAMPLES OF USE

In the American Poultry Association breed standards for poultry, for example, ducks and geese are divided by weight, and chicken breeds are divided by size. Chickens are also divided into egg laying, meat, and ornamental varieties. In cattle, breed standards allow for comparisons and the selection of the best breeds to raise.

Regulation

Some species have international governing bodies that attempt to regulate the terminology and format of breed standards internationally, but, even where such international agreement exists, not all associations for that species necessarily belong to the international governing body. The Federation Cynologique Internationale regulates breed standards for dogs internationally, but the largest dog registry, the American Kennel Club, does not belong to the international body and uses its own breed standard format.

REFERENCES

- A COMPARATIVE ANALYSIS OF ALPACA BREED TYPE AND STANDARDS, By Jude Anderson, Maggie Krieger, and Mike Safley, The Alpaca Library from Northwest Alpacas Archived April 15, 2009, at the Wayback Machine.
- Ekarius, Carol (2007). Storey's Illustrated Guide to Poultry Breeds. 210 MAS MoCA Way, North Adams MA 01247: Storey Publishing. ISBN 978-1-58017-667-5.
- Cattle breed comparisons

CHAPTER 7

Systems of Mating

Chapter Contents

INTRODUCTION

In biology, mating is the pairing of opposite-sex or hermaphroditic organisms, usually for the purposes of sexual reproduction. Some definitions limit the term to pairing between animals, while other definitions extend the term to mating in plants and fungi. Fertilization is the fusion of both sex cell or gamete. Copulation is the union of the sex organs of two sexually reproducing animals for insemination and subsequent internal fertilization. Mating may also lead to external fertilization, as seen in amphibians, fishes and plants. For the majority of species, mating is between two individuals of opposite sexes. However, for some hermaphroditic species, copulation is not required because the parent organism is capable of self-fertilization (autogamy); for example, banana slugs.

The term mating is also applied to related processes in bacteria, archaea and viruses. Mating in these cases involves the pairing of individuals, accompanied by the pairing of their homologous chromosomes and then exchange of genomic information leading to formation of recombinant progeny.

ANIMALS

For animals, mating strategies include random mating, disassortative mating, assortative mating, or a mating pool. In some birds, it includes behaviors such as nest-building and feeding offspring. The human practice of mating and artificially inseminating domesticated animals is part of animal husbandry.

In some terrestrial arthropods, including insects representing basal (primitive) phylogenetic clades, the male deposits spermatozoa on the substrate, sometimes stored within a special structure. Courtship involves inducing the female to take up the sperm package into her genital opening without actual copulation. In groups such as dragonflies and many spiders, males extrude sperm into secondary copulatory structures removed from their genital opening, which are then used to inseminate the female (in dragonflies, it is a set of modified sternites on the second abdominal segment; in spiders, it is the male pedipalps). In advanced groups of insects, the male uses its aedeagus, a structure formed from the terminal segments of the abdomen, to deposit sperm directly (though sometimes in a capsule called a «spermatophore») into the female›s reproductive tract.

Other animals reproduce sexually with external fertilization, including many basal vertebrates. Vertebrates (such as reptiles, some fish, and most birds) reproduce with internal fertilization through cloacal copulation (see also hemipenis), while mammals copulate vaginally.

- Monkeys mating

Wolves mating

Lions mating

Hermaphroditic mating snails (*Cornu aspersum*)

Mating African spurred tortoises (*Centrochelys sulcata*)

Chalkhill blue butterflies (*Polyommatus coridon*) mating

Midair mating of *Simosyrphus grandicornis*, a hoverfly

Mating pair of poplar hawk-moths (*Laothoe populi*)

Ladybugs mating

Bugs mating (*Aphrophora alni*)

PLANTS AND FUNGI

Like in animals, mating in other Eukaryotes, such as plants and fungi, denotes sexual conjugation. However, in vascular plants this is mostly achieved without physical contact between mating individuals (see pollination), and in some cases, e.g., in fungi no distinguishable male or female organs exist (see isogamy); however, mating types in some fungal species are somewhat analogous to sexual dimorphism in animals, and determine whether or not two individual isolates can mate. *Yeasts* are eukaryotic microorganisms classified in the kingdomFungi, with 1,500 species currently described. In general, under high stress conditions like nutrient starvation, haploid cells will die; under the same conditions, however, diploid cells of *Saccharomyces cerevisiae* can undergo sporulation, entering sexual reproduction (meiosis) and produce a variety of haploid spores, which can go on to mate (conjugate) and reform the diploid.

PROTISTS

Protists are a large group of diverse eukaryotic microorganisms, mainly unicellular animals and plants that do not form tissues. Eukaryotes emerged in evolution more than 1.5 billion years ago. The earliest eukaryotes were likely protists. Mating and sexual reproduction are widespread among extant eukaryotes including protists such as *Paramecium* and *Chlamydomonas*. In many eukaryotic species, mating is promoted by sex pheromones including the protist *Blepharisma japonicum.* Based on a phylogenetic analysis, Dacks and Roger proposed that facultative sex was present in the common ancestor of all eukaryotes.

However, to many biologists it seemed unlikely until recently, that mating and sex could be a primordial and fundamental characteristic of eukaryotes. A principal reason for this view was that mating and sex appeared to be lacking in certain pathogenic protists whose ancestors branched off early from the eukaryotic family tree. However, several of these protists are now known to be capable of, or to recently have had, the capability for meiosis and hence mating. To cite one example, the common intestinal parasite *Giardia intestinalis* was once considered to be a descendant of a protist lineage that predated the emergence of meiosis and sex. However, *G. intestinalis* was recently found to have a core set of genes that function in meiosis and that are widely present among sexual eukaryotes. These results suggested that *G. intestinalis* is capable of meiosis and thus mating and sexual reproduction. Furthermore, direct evidence for meiotic recombination, indicative of mating and sexual reproduction, was also found in *G. intestinalis*. Other protists for which evidence of mating and sexual reproduction has recently been described are parasitic protozoa of the genus *Leishmania*, *Trichomonas vaginalis*, and acanthamoeba.

Protists generally reproduce asexually under favorable environmental conditions, but tend to reproduce sexually under stressful conditions, such as starvation or heat shock.

REFERENCES

1. The Free Dictionary. "Mate". Retrieved May 31, 2013.
2. The Free Dictionary. "'Fertilization' - definition of". Farlex, Inc. Retrieved 25 January 2014.
3. Libbie Henrietta Hyman (15 September 1992). Hyman's Comparative Vertebrate Anatomy. University of Chicago Press. ISBN 978-0-226-87013-7.
4. "What are yeasts?". Yeast Virtual Library. 13 September 2009. Retrieved 28 November 2009.
5. Neiman, A.M. (2005). "Ascospore formation in the yeast Saccharomyces cerevisiae". Microbiology and Molecular Biology Reviews. 69 (4): 565–584.

doi:10.1128/MMBR.69.4.565-584.2005. PMC 1306807 . PMID 16339736.

6. Javaux EJ, Knoll AH, Walter MR (2001). "Morphological and ecological complexity in early eukaryotic ecosystems". Nature. 412 (6842): 66–9. doi:10.1038/35083562. PMID 11452306.
7. Dacks J, Roger AJ (1999). "The first sexual lineage and the relevance of facultative sex". J. Mol. Evol. 48 (6): 779–83. doi:10.1007/pl00013156. PMID 10229582.
8. Ramesh MA, Malik SB, Logsdon JM (2005). "A phylogenomic inventory of meiotic genes; evidence for sex in Giardia and an early eukaryotic origin of meiosis". Curr. Biol. 15 (2): 185–91. doi:10.1016/j.cub.2005.01.003. PMID 15668177.
9. Cooper MA, Adam RD, Worobey M, Sterling CR (2007). "Population genetics provides evidence for recombination in Giardia". Curr. Biol. 17 (22): 1984–8. doi:10.1016/j.cub.2007.10.020. PMID 17980591.
10. Akopyants NS, Kimblin N, Secundino N, Patrick R, Peters N, Lawyer P, Dobson DE, Beverley SM, Sacks DL (2009). "Demonstration of genetic exchange during cyclical development of Leishmania in the sand fly vector". Science. 324 (5924): 265–8. doi:10.1126/science.1169464. PMC 2729066 . PMID 19359589.
11. Malik SB, Pightling AW, Stefaniak LM, Schurko AM, Logsdon JM (2008). "An expanded inventory of conserved meiotic genes provides evidence for sex in Trichomonas vaginalis". PLoS ONE. 3 (8): e2879. doi:10.1371/journal.pone.0002879. PMC 2488364 . PMID 18663385.
12. Khan NA, Siddiqui R (2015). "Is there evidence of sexual reproduction (meiosis) in Acanthamoeba?". Pathog Glob Health. 109 (4): 193–5. doi:10.1179/2047773215Y.0000000009. PMID 25800982.

CHAPTER 8

Genotypes: Genetic Makeups of Organisms

Chapter Contents

- **Introduction**
- **Genotypic and Genomic Sequence**
- **Determining Genotype**
- **Phenotype**

Genotypes are the genetic makeups of organisms and are typically referred to with respect to the specific traits they describe. Genotypes exist in the form of genetic data such as DNA or RNA. While it is typically used to describe the genetic basis for a specific trait, the term genotype can also be used to represent the summation of a creature's genetic code. The term even applies to genetic information that is not expressed in some visible trait, as some genetic code is not actually observably expressed but is still a part of an organism's overall genetic information.

INTRODUCTION

The genotype is the part (DNA sequence) of the genetic makeup of a cell, and therefore of an organism or individual, which determines a specific characteristic (phenotype) of that cell/organism/individual. Genotype is one of three factors that determine phenotype, the other two being inherited epigenetic factors, and non-inherited environmental factors. DNA mutations which are acquired rather than inherited, such as cancer mutations, are not part of the individual's genotype; hence, scientists and physicians sometimes talk for example about the (geno)type of a particular cancer, that is the genotype of the disease as distinct from the diseased.

An example of how genotype determines a characteristic is petal color in a pea plant.

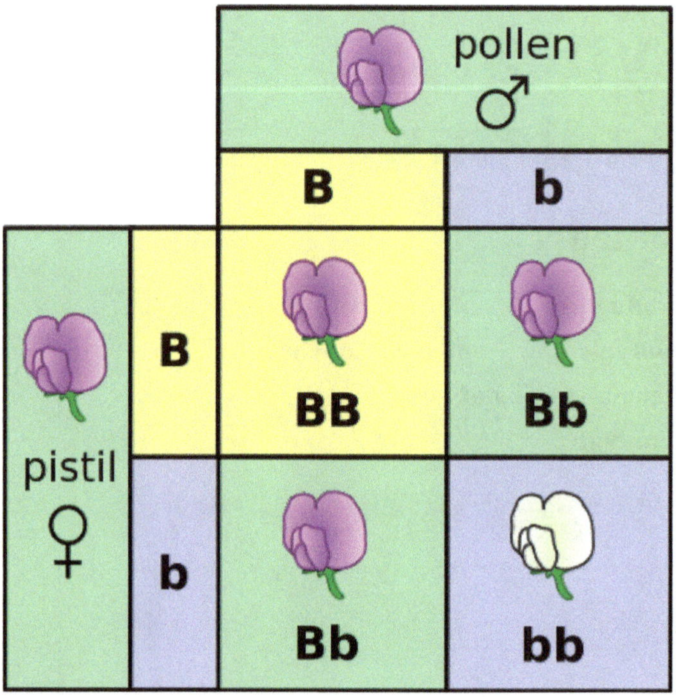

Here the relation between genotype and phenotype is illustrated, using a Punnett square, for the character of petal colour in a pea plant. The letters B and b represent genes for colour and the pictures show the resultant flowers.

GENOTYPIC AND GENOMIC SEQUENCE

The genotype of an organism is the inherited map it carries within its genetic code. Not all organisms with the same genotype look or act the same way because appearance and behavior are modified by environmental and developmental conditions. Likewise, not all organisms that look alike necessarily have the same genotype. One's genotype differs subtly from one's genomic sequence. A sequence is an absolute measure of base composition of an individual, or a representative of a species or group; a genotype typically implies a measurement of how an individual *differs* or is specialized within a group of individuals or a species. So typically, one refers to an individual's genotype with regard to a particular gene of interest and, in polyploid individuals, it refers to what combination of alleles the individual carries (see homozygous, heterozygous). The genetic constitution of an organism is referred to as its genotype, such as the letters Bb. (B - dominant genotype and b - recessive genotype)

Genotype and Phenotype

Any given gene will usually cause an observable change in an organism, known as the phenotype. The terms genotype and phenotype are distinct for at least two reasons:

- To distinguish the source of an observer's knowledge (one can know about genotype by observing DNA; one can know about phenotype by observing outward appearance of an organism).
- Genotype and phenotype are not always directly correlated. Some genes only express a given phenotype in certain environmental conditions. Conversely, some phenotypes could be the result of multiple genotypes. The genotype is commonly mixed up with the phenotype which describes the end result of both the genetic and the environmental factors giving the observed expression (e.g. blue eyes, hair color, or various hereditary diseases).

A simple example to illustrate genotype as distinct from phenotype is the flower colour in pea plants (see Gregor Mendel). There are three available genotypes, PP (homozygous dominant), Pp (heterozygous), and pp (homozygous recessive). All three have different genotypes but the first two have the same phenotype (purple) as distinct from the third (white).

A more technical example to illustrate genotype is the single nucleotide polymorphism or SNP. A SNP occurs when corresponding sequences of DNA from different individuals differ at one DNA base, for example where the sequence AAGCCTA changes to AAGCTTA. This contains two alleles : C and T. SNPs typically have three genotypes, denoted generically AA Aa and aa. In the example above, the three genotypes would be CC, CT and TT. Other types of genetic marker, such as microsatellites, can have more than two alleles, and thus many different genotypes.

Genotype and Mendelian Inheritance

The distinaction between genotype and phenotype is commonly experienced when studying family patterns for certain hereditary diseases or conditions, for example, haemophilia. Due to the diploidy of humans (and most animals), there are two alleles for any given gene. These alleles can be the same (homozygous) or different (heterozygous), depending on the individual (see zygote). With a dominant allele, the offspring is guaranteed to inherit the trait in question irrespective of the second allele.

In the case of an albino with a recessive allele (aa), the phenotype depends upon the other allele (Aa, aA, aa or AA). An affected person mating with a heterozygous individual (Aa or aA, also carrier) there is a 50-50 chance the offspring will be albino's phenotype. If a heterozygote mates with another heterozygote, there is 75% chance passing the gene on and only a 25% chance that the gene will be displayed. A homozygous dominant (AA) individual has a normal phenotype and no risk of abnormal offspring. A homozygous recessive individual has an abnormal phenotype and is guaranteed to pass the abnormal gene onto offspring.

In the case of haemophilia, it is sex-linked thus only carried on the X chromosome. Only females can be a carrier in which the abnormality is not displayed. This woman has a normal phenotype, but runs a 50-50 chance, with an unaffected partner, of passing her abnormal gene on to her offspring. If she mated with a man with haemophilia (another carrier) there would be a 75% chance of passing on the gene.

DETERMINING GENOTYPE

Genotyping is the process of elucidating the genotype of an individual with a biological assay. Also known as a *genotypic assay*, techniques include PCR, DNA fragment analysis, allele specific oligonucleotide (ASO) probes, DNA sequencing, and nucleic acid hybridization to DNA microarrays or beads. Several common genotyping techniques include restriction fragment length polymorphism (*RFLP*), terminal restriction fragment length polymorphism (*t-RFLP*), amplified fragment length polymorphism (*AFLP*), and multiplex ligation-dependent probe amplification (*MLPA*).

DNA fragment analysis can also be used to determine such disease causing genetics aberrations as microsatellite instability (*MSI*), *trisomy* or aneuploidy, and loss of heterozygosity (*LOH*). MSI and LOH in particular have been associated with cancer cell genotypes for colon, breast and cervical cancer.

The most common chromosomal aneuploidy is a trisomy of chromosome 21 which manifests itself as Down syndrome. Current technological limitations typically allow only a fraction of an individual's genotype to be determined efficiently.

PHENOTYPE

A **phenotype** (from Greek *phainein*, meaning "to show", and *typos*, meaning "type") is the composite of an organism's observable characteristics or traits, such as its morphology, development, biochemical or physiological properties, behavior, and products of behavior (such as a bird's nest). A phenotype results from the expression of an organism's genetic code, its genotype, as well as the influence of environmental factors and the interactions between the two. When two or more clearly different phenotypes exist in the same population of a species, the species is called polymorphic.

This genotype-phenotype distinction was proposed by Wilhelm Johannsen in 1911 to make clear the difference between an organism's heredity and what that heredity produces. The distinction is similar to that proposed by August Weismann, who distinguished between germ plasm (heredity) and somatic cells (the body). The genotype-phenotype distinction should not be confused with Francis Crick's central dogma of molecular biology, which is a statement about the directionality of molecular sequential information flowing from DNA to protein, and not the reverse.

Richard Dawkins in 1978 and then again in his 1982 book *The Extended Phenotype* suggested that bird nests and other built structures such as caddis fly larvae cases and beaver dams can be considered as "extended phenotypes".

The shells of individuals within the bivalve mollusk species Donax variabilis show diverse coloration and patterning in their phenotypes

Difficulties in Definition

Despite its seemingly straightforward definition, the concept of the phenotype has hidden subtleties. It may seem that anything dependent on the genotype is a phenotype, including molecules such as RNA and proteins. Most molecules and structures coded by the genetic material are not visible in the appearance of an organism, yet they are observable (for example by Western blotting) and are thus part of the phenotype.

Human blood groups are an example. It may also seem that this goes beyond the original intentions of the concept with its focus on the (living) organism in itself, meaning that the lowest level of biological organization compatible with the phenotype concept is at the cellular level. Either way, the term phenotype includes traits or characteristics that can be made visible by some technical procedure. A notable extension to this idea is the presence of "organic molecules" or metabolites that are generated by organisms from chemical reactions of enzymes, e.g. vitamins, that can be scored as phenotype.

Another extension adds behavior to the phenotype, since behaviors are also observable characteristics. Indeed, there is research into the clinical relevance of behavioral phenotypes as they pertain to a range of syndromes. The term "phenotype" has sometimes been incorrectly used as a shorthand for phenotypic difference from wild type, bringing the absurd statement that a mutation has no phenotype.

Biston betularia morpha *typica*, the standard light-colored Peppered Moth.

Biston betularia morpha *carbonaria*, the melanic Peppered Moth, illustrating discontinuous variation.

Phenotypic Variation

Phenotypic variation (due to underlying heritable genetic variation) is a fundamental prerequisite for evolution by natural selection. It is the living organism as a whole that contributes (or not) to the next generation, so natural selection affects the genetic structure of a population indirectly via the contribution of phenotypes. Without phenotypic variation, there would be no evolution by natural selection.

The interaction between genotype and phenotype has often been conceptualized by the following relationship:

genotype (G) + environment (E) → phenotype (P)

A more nuanced version of the relationship is:

genotype (G) + environment (E) + genotype & environment interactions (GE) → phenotype (P)

Genotypes often have much flexibility in the modification and expression of phenotypes; in many organisms these phenotypes are very different under varying environmental conditions (see ecophenotypic variation). The plant *Hieracium umbellatum* is found growing in two different habitats in Sweden. One habitat is rocky, sea-side cliffs, where the plants are bushy with broad leaves and expanded inflorescences; the other is among sand dunes where the plants grow prostrate with narrow leaves and compact inflorescences. These habitats alternate along the coast of Sweden and the habitat that the seeds of *Hieracium umbellatum* land in, determine the phenotype that grows.

An example of random variation in *Drosophila* flies is the number of ommatidia, which may vary (randomly) between left and right eyes in a single individual as much as they do between different genotypes overall, or between clones raised in different environments.

The concept of phenotype can be extended to variations below the level of the gene that affect an organism›s fitness. For example, silent mutations that do not change the corresponding amino acid sequence of a gene may change the frequency of guanine-cytosine base pairs (GC content). These base pairs have a higher thermal stability (*melting point*, see also DNA-DNA hybridization) than adenine-thymine, a property that might convey, among organisms living in high-temperature environments, a selective advantage on variants enriched in GC content.

The Extended Phenotype

The idea of the phenotype has been generalized by Richard Dawkins in *The Extended Phenotype* to mean all the effects a gene has on the outside world that may influence its chances of being replicated. These can be effects on the organism in which the gene resides, the environment, or other organisms. The term "extended phenotype" was first coined in 1978.

For instance, a beaver dam might be considered a phenotype of beaver genes, the same way beavers› powerful incisor teeth are phenotype expressions of their genes. Dawkins also cites the effect of an organism on the behavior of another organism (such as the devoted nurturing of a cuckoo by a parent of a different species) as an example of the extended phenotype as well as parasites living inside the body of a host. The first example he used was sporocysts of flukes of the genus *Leucochloridium* that invade the tentacles of snails where they can be seen conspicuously pulsating through the snail's skin.

This change in both color and behavior (infected snails move upwards on vegetation) is suggested to increase predation on the snail by birds and therefore to assist the parasite enter its final host, a bird. The third example of the extended phenotype is «Action at a Distance». This is where genes in one organism affect the behavior of another organism. The examples Dawkins used were genes in orchids affecting orchid bee behavior (to increase pollination), genes in rattlesnakes causing avoidance behavior in other animals, and genes in male peacocks affecting copulatory decisions of peahens.

The smallest unit of replicators is the gene. Replicators cannot be directly selected upon, but they are selected on by their phenotypic effects. These effects are packaged together in organisms. We should think of the replicator as having extended phenotypic effects. These are all of the ways it affects the world, not just the effects the replicators have on the body in which they reside.

Phenome and Phenomics

Although a phenotype is the ensemble of observable characteristics displayed by an organism, the word *phenome* is sometimes used to refer to a collection of traits, while the simultaneous study of such a collection is referred to as *phenomics*.

REFERENCES

1. "Genotype definition - Medical Dictionary definitions". medterms.com.
2. Hulce, David; Liu, ChangSheng (Jonathan) (July 2006). "SoftGenetics Application Note - GeneMarker® Software for Terminal-Restriction Fragment Length Polymorphism (T-RFLP) Data Analysis" (PDF). SoftGenetics.
3. "Keygene.com Homepage"
4. "SoftGenetics Application Note - Software for Multiplex Ligation-dependent Probe Amplification (MLPA™)" (PDF). SoftGenetics. April 2006. Retrieved 2011-03-13.
5. He, Haiguo; Ning, Wan; Liu, Jonathan (March 2007). «SoftGenetics Application Note - Microsatellite Instability Analysis with GeneMarker® Tamela Serensits» (PDF). SoftGenetics.
6. "SoftGenetics Application Note - GeneMarker® Software for Trisomy Analysis" (PDF). SoftGenetics. November 2006.
7. Serensits, Pamela; He, Haiguo, Ning, Wan; Liu, Jonathan (March 2007). "SoftGenetics Application Note - Loss of Heterozygosity Detection with GeneMarker" (PDF). SoftGenetics.

CHAPTER 9

Cuniculture

Chapter Contents

- **Introduction**
- **Aspects of Rabbit Production**
- **Husbandry**
- **Exhibition and Fancier Societies**

The rabbit breeding is the breeding of domestic rabbits . It developed from the Middle Ages in Europe, but has really taken off in the world recently. Its main purpose is the production of meat , and sometimes the hair (angora rabbits) or fur or laboratory animal supply for their experiments. It is also a breeding of leisure for some, which present their pure-bred animals during fairs and exhibitions. Recently, the rabbit is as high as pet .

INTRODUCTION

Cuniculture is the agricultural practice of breeding and raising domestic rabbits, usually for their meat, fur, or wool. This differs from the simpler practice of keeping a single or small group of rabbits as companions, without selective breeding, reproduction, or the care of young animals. Some people, called *rabbit fanciers*, practice cuniculture predominantly for exhibition. The distribution of rabbit farming varies around the globe, and while it is on the decline in some nations, in others it is expanding.

Rabbits kept in cages for scientific experimentation

Early Husbandry

Phoenician sailors visiting the coast of Spain c. 12th century BC, mistaking the European rabbit for a species from their homeland (the rock hyrax *Procavia capensis*), gave it the name *i-shepan-ham* (land or island of hyraxes). A theory exists that a corruption of this name, used by the Romans, became the Latin name for Spain, Hispania – although this theory is somewhat controversial.

Domestication of the European rabbit rose slowly from a combination of game-keeping and animal husbandry. Among the numerous foodstuffs imported by sea to Rome during her domination of the Mediterranean were shipments of rabbits from Spain. Romans also imported ferrets for rabbit hunting, and the Romans then distributed rabbits and the habit of rabbit keeping to the rest of Italy, to France, and then across the Roman Empire, including the British Isles. Rabbits were kept in both walled areas as well as more extensively in game-preserves. In the British Isles, these preserves were known as warrens or garths, and rabbits were known as coneys, to differentiate them from the similar hares (a separate species). The term warren was also used as a name for the location where hares, partridges and pheasants were kept, under the watch of a game keeper called a warrener. In order to confine and protect the rabbits, a wall or thick hedge might be constructed around the warren, or a warren might be established on an island. (see: Rabbit islands) A warrener was responsible for controlling poachers and other predators and would collect the rabbits with snares, nets, hounds (such as greyhounds), or by hunting with ferrets. With the rise of falconry, hawks and falcons were also used to collect rabbits and hares.

Domestication

While under the warren system, rabbits were managed and harvested, they were not domesticated. The practice of rabbit domestication also came from Rome. Christian monasteries throughout Europe and the Middle East kept rabbits since at least the 5th century. (Pope Gregory stated in a Papal Edict of the year 600 AD that fetal rabbits were permissible to eat during the Lenten fast, greatly enhancing their popularity, and it is from this date that the true domestication of rabbits is counted.) While rabbits might be allowed to wander freely within the monastery walls, a more common method was the employment of rabbit courts or rabbit pits. A rabbit court was a walled area lined with brick and cement, while a pit was much the same, only less well-lined, and more sunken. Individual boxes or burrow-spaces could line the wall. Rabbits would be kept in a group in these pits or courts, and individuals collected when desired for eating or pelts. From these pits, which did not allow for easy cleaning, ready handling of rabbits, or for selective breeding, rabbit keepers transitioned to individual hutches or pens, which were originally made of wood but are now more frequently made of metal in order to allow for better sanitation.

Early Breeds

Rabbits were typically kept as part of the household livestock by peasants and villagers throughout Europe. Husbandry of the rabbits, including collecting weeds and grasses for fodder, typically fell to the children of the household or farmstead. These rabbits were largely 'common' or 'meat' rabbits and not of a particular breed, although regional strains and types did arise. Some of these strains remain as regional breeds, such as the Gothland of Sweden, while others, such as the Land Kaninchen, a spotted rabbit of Germany, have become extinct. Another rabbit type that standardized into a breed was the Brabancon, a meat rabbit of the region of Limbourg and what is now Belgium. Rabbits of this breed were bred for the Ostend port market, destined for London markets. (Whitman, pg 10) The development of the refrigerated shipping vessels led to the eventual collapse of the European meat rabbit trade, as the over-population of feral rabbits in Australia could now be harvested and sold. The Brabancon is now considered extinct, although a descendant, the Dutch breed, remains a popular small rabbit for the pet trade.

In addition to being harvested for meat, properly prepared rabbit pelts were also an economic factor. Both wild rabbits and domestic rabbit pelts were valued, and it followed that pelts of particular rabbits would be more highly prized. As far back as 1631, price differentials were noted between ordinary rabbit pelts and the pelts of quality 'riche' rabbit in the Champagne region of France. (This regional type would go on to be recognized as the *Champagne D'Argent*, the *silver rabbit of Champagne*.)

Among the earliest of the commercial breeds was the Angora, which some say may have developed in the Carpatian mountains. They made their way to England, where during the rule of King Henry VIII, laws banned the exportation of long-haired rabbits as a national treasure. In 1723, long haired rabbits were imported to southern France by English sailors, who described the animals as originally coming from the Angora region of Turkey. Thus two distinct strains arose, one in France and one in England.

Expansion Around the Globe

European explorers and sailors took rabbits with them to new ports around the world, and brought new varieties back to Europe and England with them. With the second voyage of Christopher Columbus in 1494, European domestic livestock were brought to the New World. Rabbits, along with goats and other hardy livestock, were frequently released on islands to produce a food supply for later ships. The importations occasionally met with disastrous results. (See Rabbits in Australia) While cattle and horses were used across the socio-economic spectrum, and especially were concentrated among the wealthy, rabbits were kept by lower-income classes and peasants. This is reflected in the names given to the breeds that eventually arose in the colonized areas. From the Santa Duromo mountains of Brazil comes the Rustico, which is known in the United States as the Brazilian rabbit.) The Criollo rabbit comes from Mexico.

International Commercial Use

With the rise of scientific animal breeding in the late 1700s, led by Robert Bakewell (among others), distinct livestock breeds were developed for specific purposes. Rabbits were among the last of the domestic animals to have these principles applied to them, but the rabbit's rapid reproductive cycle allowed for marked progress towards a breeding goal in a short period of time. Additionally, rabbits could be kept on a small area, with a single person caring for over 300 breeding does on an acre of land. Rabbit breeds were developed by individuals, cooperatives, and by national breeding centers. To meet various production goals, rabbits were exported around the world. One of the most notable import events was the introduction of the Belgium Hare breed of rabbit from Europe to the United States. This led to a short-lived "boom" in rabbit breeding, selling, and speculation, when a quality breeding animal could bring \$75 to \$200. (For comparison, the average daily wage was approximately \$1.00.) In 1900, a single animal export company recorded 6,000 rabbits successfully shipped to the United States and Canada.

Science played another role in rabbit raising – this time with rabbits themselves as the tools used for scientific advancement. Beginning with Louis Pasteur's experiments in rabies in the later half of the nineteenth century, rabbits have been used as models to investigate various medical and biological problems, including the transmission of disease and protective antiserums. Production of quality animals for meat sale and scientific experimentation has driven a number of advancements in rabbit husbandry and nutrition. While early rabbit keepers were limited to local & seasonal foodstuffs, which did not permit the maximization of production, health or growth, by 1930 researchers were conducting experiments in rabbit nutrition, similar to the experiments that had isolated vitamins and other nutritional components. This eventually resulted in the development of various recipes for pelleted rabbit diets. Gradual refinement of diets has resulted in the widespread availability of pelleted diets which increase yield, reduce waste, and promote rabbit health, particularly maternal breeding health.

Rise of the Fancy

The final leg of deliberate rabbit breeding — beyond meat, wool & fur, and laboratory use — was the breeding of 'fancy' animals as pets and curiosity. The term 'fancy' was originally applied to long eared 'lop' rabbits, as the lop rabbits were the first rabbits bred for exhibition. They were first admitted to agricultural shows in England in the 1820s, and in 1840 a club was formed for the promotion and regulation of exhibitions for "Fancy Rabbits". In 1918, a new group formed for the promotion of fur breeds, originally including only Beverans and Havana breeds. This club eventually expanded into the British Rabbit Council. Meanwhile, in the United States, clubs promoting various breeds were chartered in the 1880s, and the National Pet Stock Association was formed in 1987. This organization would become the American Rabbit Association. Many thousand rabbit shows occur each year and are sanctioned in Canada, Mexico, Malaysia, Indonesia and the United States by the ARBA.

With the advent of national organizations, rabbit breeders had a framework for establishing breeds and varieties against recognized standards. Thus rabbit exhibition breeding began to rapidly expand. Organizations and associations were also established across Europe, most notably in Germany, France, and Scandinavia, allowing for the recognition of local breeds (many of which shared similar characteristics across national borders) and for the preservation of stock during such disruptions as World War I and World War II.

Closely overlapping with expedition breeding and fur breeding has been the breeding of rabbits for the pet trade. While accurate production records are not readily available, rabbits have been kept as pets for centuries. The sale of rabbits as pets rose in the last half of the twentieth century, as rabbits could be kept in smaller living areas than more traditional companion animals such as dogs and cats, but did not require as specialized housing as gold fish. Several strains of rabbit – such as the Holland Lop, the Polish rabbit, the Netherland Dwarf, and the Lionhead – have been specifically bred for the pet trade. Traits common to many popular pet breeds are small size, "dwarf" features, and patterned coats.

Modern Farming

Outside of the exhibition circles, rabbit raising remained a small scale but persistent household and farm endeavor, in many locations unregulated by the rules that governed the production of larger livestock. With the ongoing urbanization of world-wide population, rabbit raising gradually declined, but saw resurgences in both Europe and North America during World War II, in conjunction with victory gardens. Eventually, farmers across Europe and in the United States began to approach cuniculture with the same scientific principles as had already been applied to the production of grains, poultry, and hoofstock. National agriculture breeding stations were established to improve local rabbit strains and to introduce more productive breeds. National breeding centers focused most on developing strains for production purposes, including meat, pelts, and wool. These gradually faded from prominence in the United States, but remained viable longer in Europe. Meanwhile, rabbit raising for local markets gained prominence in developing nations as an economical means of producing protein. Various aid agencies promote the use of rabbits as livestock. The animals are particularly useful in areas where women are limited in employment outside the household, because rabbits can be kept successfully in small areas. These same factors have contributed to the increased popularity of rabbits as 'backyard livestock' among locavores and homesteaders in more developed countries in North America and Europe. The addition of rabbit breeds to the listing of endangered heritage breeds by the American Livestock Breeds Conservatory has also led to increased interest for livestock conservationists. In contrast, throughout Asia, and particularly in China, rabbits are increasingly raised intensively and sold for export around the world.

In addition to the previously mentioned national ministries and associations, the World Rabbit Science Association was formed in 1976, as a venue for distributing the most current information in rabbit psyology, medicine, and husbandry. The association puts on a conference every four years, most recently 2012 in Egypt.

Present Day

In more recent years and in some countries, cuniculture has come under pressure from animal rights activists on several fronts. The use of animals, including rabbits, in scientific experiments has been subject to increased scrutiny in developed countries. Increasing regulation has raised the cost of producing animals for this purpose, and made other experimental options more attractive. Other researchers have abandoned investigations which required animal models. Meanwhile, various rescue groups under the House Rabbit Society umbrella have taken an increasingly strident stance against any breeding of rabbits (even as food in developing countries) on the grounds that it contributes to the number of mistreated, unwanted or abandoned animals. Some of these organizations have promoted investigation and prosecution of rabbit raisers on humanitarian concerns. Some rabbit raisers have protested these investigations as being biased and conducted on illegitimate grounds. Finally, the growth of homesteaders and small holders has led to the rise of visibility of rabbit raisers in geographic areas where they have not been present previously. This has led to zoning conflicts over the regulation of butchering and waste management. Conflicts have also arisen with House Rabbit Society organizations as well as ethical vegetarians and vegans concerning the use of rabbits as meat and fur animals rather than as pets. Ironically, many homesteaders cite concern with animal welfare in intensive farming of beef, pork and poultry as a significant factor in choosing to raise rabbits for meat.

The specific future direction of cuniculture is unclear, but does not appear to be in danger of disappearing in any particular part of the world. The variety of applications, as well as the versatile utility of the species, appears sufficient to keep rabbit raising a going concern in one aspect or another around the planet.

ASPECTS OF RABBIT PRODUCTION

Meat Rabbits

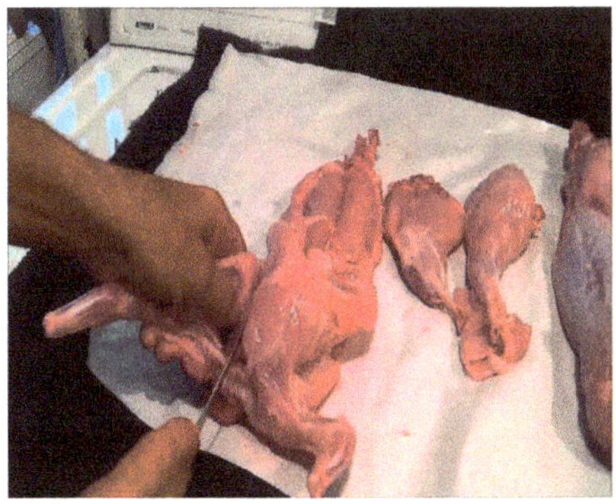

Commercially processed lean rabbit meat

Meat-type rabbits being raised as a supplementary food source during the Great Depression

Rabbits have been raised for meat production in a variety of settings around the world. Small-scale smallholder or backyard operations remain common in many countries, while larger scale commercial operations occur in Europe and Asia. For smaller operations, local breeds of various types may be used. Many local, 'rustico', landrace or other heritage type breeds may be used only in a specific geographic area. Sub-par or cull animals from other breeding goals (laboratory, exhibition/show, wool, pet) may also be used for meat (particularly in smallholder operations). However, dwarf breeds are rarely primarily used for meat production, due to the small size, slow growth, and low numbers of offspring per litter. Likewise, the giant breeds (fourteen pounds and up at adult size) are not commonly used for meat production, due to extended growth rates that lead to high feed costs and large bone size that reduces the dress-out percentage.

In contrast to the multitude of breeds & types used in smaller operations, breeds such as the New Zealand and Californian, as well as hybrids of these breeds, are most frequently utilized for meat in commercial rabbitries. The primary qualities of good meat rabbit breeding stock are growth rate and mothering ability. Uniform growth rates & size at slaughter are also considered important factors. Specific lines of commercial breeds have been developed that maximize these qualities - rabbits may be slaughtered as early as seven weeks and does of these strains routinely raise litters of 8 to 12 kits. Other breeds of rabbit developed for commercial meat production include the Florida White and the Altex.

Rabbit breeding stock raised in France is particularly popular with meat rabbit farmers internationally, some being purchased as far away as in China in order to improve the local rabbit herd.

Larger-scale operations attempt to maximize income by balancing land use, labor involved, animal health, and investment in infrastructure. Specific infrastructure and strain qualities depend on the geographic area. An operation in an urban area may emphasize odor control and space utilization by stacking cages over each other with automatic cleaning systems that flush away feces and urine. In rural sub-tropical and tropical areas, temperature control becomes more of an issue, and the use of air-conditioned buildings is common in many areas.

Breeding schedules for rabbits vary by individual operation. Prior to the development of modern balanced rabbit rations, rabbit breeding was limited by the nutrition available to the doe. Without adequate calories and protein, the doe would either not be fertile, would abort or re-adsorb the foetuses during pregnancy, or would deliver small numbers of weak kits. Under these conditions, a doe would be re-bred only after weaning her last litter when the kits reached the age of two months. This allowed for a maximum of four litters per year. Advances in nutrition, such as those published by the USDA Rabbit Research Station, resulted in greater health for breeding animals and the survival of young stock. Likewise, offering superior, balanced nutrition to growing kits allowed for better health and less illness among slaughter animals. Current practices include the option of re-breeding the doe within a few days of delivery (closely matching the behavior of wild rabbits during the spring/early summer, when forage availability is at its peak.) This can result in up to eight or more litters annually. A doe of ideal meat-stock genetics can produce five times her body weight in fryers a year. Criticism of the more intensive breeding schedules has been made, on the grounds that re-breeding that closely is excessively stressful for the doe. Determination of health effects of breeding schedules is made more difficult by the domestic rabbit's reproductive psychology - in contrast to several other mammal species, rabbits are more likely to develop uterine cancer when not used for breeding than when bred frequently.

Rabbit fryers are rabbits that are between 70 and 90 days of age, and weighing between 3 and 5 lb (1 to 2 kg) live weight. Rabbit roasters are rabbits from 90 days to 6 months of age weighing between 5 and 8 lb (2 to 3.5 kg) live weight. Rabbit stewers are rabbits from 6 months on weighing over 8 lb. Dark fryers (any other color but whites) are typically lower in price than albino fryers because of the slightly darker tinge of the fryer (purely pink carcasses are preferred by consumers) and because the hide is harder to remove manually than the white albino fryers.

Commercial The highest prices per pound of live weight in the United States are offered for fryers. In Europe, however, a sizable market remains for the larger & older rabbits. As the name implies, rabbit meat that comes from older animals is cooked differently from that of young fryer rabbits.

By some estimates, world's annual rabbit meat production stood at around 1.5 million tons in 1990. In 2014, the number was reported by some sources as at around 2 million tons.China is among the world's largest producers and consumers, accounting for some 30% of the world's total rabbit meat consumption. Within China itself, rabbits are raised in many provinces, most of it (about 70% of the national production, i.e. some 420,000 tons annually) being consumed in the Sichuan Basin (Sichuan Province and Chongqing), where it is particularly popular.

Well-known chef Mark Bittman says that domesticated rabbit tastes like chicken because both are blank palettes upon which any desired flavors can be layered.

Wool Rabbits and Pelt Rabbits

Wool Rabbits

Rabbits such as the Angora, American Fuzzy Lop, and Jersey Wooly produce wool. However, since the American Fuzzy Lop and Jersey Wooly are both dwarf breeds, only the much larger Angora breeds such as the English Angora, Satin Angora, Giant Angora, and French Angoras are used for commercial wool production. Their long fur is sheared, combed, or plucked (gently pulling loose hairs from the body during molting) and then spun into yarn used to make a variety of products. Angora sweaters can be purchased in many clothing stores and is generally mixed with other types of wool. In 2010, 70% of Angora rabbit wool was produced in China. Rabbit wool, called Angora, is 5 times warmer than sheep's wool.

Fur Rabbits

Dried rabbit pelts

All rabbits produce fur. Rabbits such as the Palomino, Satin, Chinchilla rabbit and Rex rabbit are commonly raised for fur. Each breed has unique coloring and fur characteristics. The rabbit is fed a diet especially balanced for fur production and the pelts are harvested when they have reached prime condition. Rabbit fur is widely used throughout the world. China imports much of its fur from Scandinavia (80%) and North America (5%) according to the USDA Foreign Agricultural Service GAIN Report CH7607.

Exhibition Rabbits

Many rabbit keepers breed their rabbits for competition among other purebred rabbits of the same breed. Rabbits are judged according to the standards put forth by the governing associations of the particular country. These associations, being made up of people, may be distinctly political and reflect the preferences of particular persons on the governing boards. However, as mechanisms to preserve rare breeds of rabbits, foster communication between breeders and encourage the education of the public, these organizations are invaluable. Examples include the American Rabbit Breeders Association and the British Rabbit Council.

Laboratory Rabbits

Rabbits have been and continue to be used in laboratory work such as production of antibodies for vaccines and research of human male reproductive system toxicology. Experiments with rabbits date back to Louis Pausture's work in France in the 1800s. In 1972, around 450 000 rabbits were used for experiments in the United States, decreasing to around 240 000 in 2006. The Environmental Health Perspective, published by the National Institute of Health, states, "The rabbit is an extremely valuable model for studying the effects of chemicals or other stimuli on the male reproductive system." According to the Humane Society of the United States, rabbits are also used extensively in the study of bronchial asthma, stroke prevention treatments, cystic fibrosis, diabetes, and cancer.

Rabbit cultivation intersects with research in two ways: first, the keeping and raising of animals for testing of scientific principles. Some experiments require the keeping of several generations of animals treated with a particular drug, in order to fully appreciate the side effects of that drug. There is also the matter of breeding and raising animals for experiments. The New Zealand White is one of the most commonly used breeds for research and testing. Specific strains of the NZW have been developed, with differing resistance to disease and cancers. Additionally, some experiments call for the use of 'specific pathogen free' animals, which require specific husbandry and intensive hygiene.

Animal rights activists generally oppose animal experimentation for all purposes, and rabbits are no exception. The use of rabbits for the Draize test, which is used for, amongst other things, testing cosmetics on animals, has been cited as an example of cruelty in animal research. Albino rabbits are typically used in the Draize tests because they have less tear flow than other animals and the lack of eye pigment make the effects easier to visualize. Rabbits in captivity are uniquely subject to rabbitpox, a condition that has not been observed in the wild.

HUSBANDRY

Rabbits being raised on pasture at Polyface Farm

Modern methods for housing domestic rabbits vary from region to region around the globe and by type of rabbit, technological & financial opportunities and constraints, intended use, number of animals kept, and the particular preferences of the owner/farmer. Various goals include maximizing number of animals per land unit (especially common in areas with high land values or small living areas) minimizing labor, reducing cost, increasing survival and health of animals, and meeting specific market requirements (such as for clean wool, or rabbits raised on pasture.) Not all of these goals are complementary. Where the keeping of rabbits has been regulated by governments, specific requirements have been put in place. Various industries also have commonly accepted practices which produce predictable results for that type of rabbit product.

Extensive Cuniculture Practices

Extensive cuniculture refers to the practice of keeping rabbits at a lower density and a lower production level than intensive culture. Specifically as relates to rabbits, this type of production was nearly universal prior to germ theory understanding of infectious parasites (especially coccidia) and the role of nutrition in prevention of abortion and reproductive loss. The most extensive rabbit "keeping" methods would be the harvest of wild or feral rabbits for meat or fur market, such as occurred in Australia prior to the 1990s. Warren-based cuniculture is somewhat more controlled, as the animals are generally kept to a specific area and a limited amount of supplemental feeding provided. Finally, various methods of raising rabbits with pasture as the primary food source have been developed. Pasturing rabbits within a fence (but not a cage) also known as colony husbandry, has not been commonly pursued due to the high death rate from weather and predators. More commonly (but still rare in terms of absolute numbers of rabbits and practitioners) is the practice of confining the rabbits to a moveable cage with an open or slatted floor so that the rabbits can access grass but still be kept at hand and protected from weather and predators. This method of growing rabbits does not typically result in an over-all reduction for the need for supplemented feed. The growing period to market weight is much longer for grass fed rather than pellet fed animals, and many producers continue to offer small amounts of complete rations over the course of the growing period. Hutches or cages for this type of husbandry are generally made of a combination of wood and metal wire, made portable enough for a person to move the rabbits daily to fresh ground, and of a size to hold a litter of 6 to 12 rabbits at the market weight of 4 to 5 pounds. Protection from sun and driving rain are important health concerns, as is durability against predator attacks and the ability to be cleaned to prevent loss from coccidious. Medical care and the use of medicated feed are less common.

Intensive Cuniculture Practices

Intensive cuniculture involves a greater density of animals per unit of land and a generally higher rate of reproduction. Labor required for each hide, kilogram of wool or market fryer may be higher or lower than for extensive methods. The amount of supplemented feed is generally higher, and may be grain mixes or a complete pelleted feed. The total number of animals may be in the thousands or may be less than ten adult animals, depending on the particular operation. Housing may be highly climate controlled (inside housing) or nearly completely outside, with minimal temperature control. Outdoor housing may consist of a single or group of cages, hutches, or lined pits. One example of intensive rabbit keeping may be a household rabbitry of a buck and two does, set under protection from the sun and with a windbreak, where each rabbit had their own cage and are fed fodder that is gathered by family members from the yard or roadside. Another might be a barn that is completely enclosed and climate controlled, to include constant air exchange, where several hundred individually housed does are fed a complete pelleted ration and are subject to weekly weight checks and daily health inspections. While a small rabbitry may make use of solid floors, wire or slat floors are more common in larger barns in order to ensure adequate hygiene. A higher level of routine preventative medical care and a more strict application of biosecurity principles are more common in intensive cuniculture, but the level of clinical intervention on individual animals is unlikely to be high.

Challenges to Successful Production

Specific challenges to the keeping of rabbits vary by specific practices. Losses from coccidiosis are much more common when rabbits are kept on the ground (such as in warrens or colonies) or on solid floors than when on wire or slat cages that keep rabbits elevated away from urine and faeces. Pastured rabbits are more subject to predator attack. Rabbits kept indoors at an appropriate temperature rarely suffer heat loss in comparison to rabbits housed outdoors in summer. At the same time, if rabbits are housed inside without adequate ventilation, respiratory disease can be a significant cause of disease and death. Production does on fodder are rarely able to raise more than 3 litters a year without heavy losses from deaths of weak kits, abortion, and re-adsorption, all related to poor nutrition and inadequate protein intake. In contrast, rabbits fed commercial pelleted diets can face losses related to low fiber intake.

EXHIBITION AND FANCIER SOCIETIES

Checkered Giant at an exhibition

In the early 1900s, as animal fancy in general began to emerge, rabbit fanciers began to sponsor rabbit exhibitions and fairs in Western Europe and the United States. What became known as the "Belgian Hare Boom", began with the importation of the first Belgian Hares from England in 1888 and soon after the founding of the first rabbit club in America, the American Belgian Hare Association. From 1898 to 1901, many thousands of Belgian Hares were imported to America. Today, the Belgian Hare is considered one of the rarest breeds with less than 200 in the United States as reported in a recent survey.

The American Rabbit Breeders Association (ARBA) was founded in 1910 and is the national authority on rabbit raising and rabbit breeds having a uniform Standard of Perfection, registration and judging system.

Conformation Shows

Show rabbits are an increasingly popular activity. Showing rabbits helps to improve the vigor and physical behavior of each breed through competitive selection. County fairs are common venues through which rabbits are shown in the United States. Rabbit clubs at local state and national levels hold many shows each year. On any given weekend one may be able to find a show in most regions of the United States and the United Kingdom. Although only purebred animals are shown, a pedigree is not required to enter a rabbit in an ARBA-sanctioned show but is required to register your rabbit with the American Rabbit Breeders Association (ARBA). A rabbit must be registered in order to receive a Grand Champion certificate. Children's clubs such as 4-H also include rabbit shows, usually in conjunction with county fairs. The ARBA holds an annual national convention which has as many as 25,000 animals competing from all over the world. The mega show moves to a different city each year. The ARBA also sponsors youth programs for families as well as underprivileged rural and inner city children to learn responsible care and breeding of domestic rabbits.

Genetics

There are 10 color gene groups (or loci) in rabbits. They are A, B, C, D, E, En, Du, Si, V, and W. Each locus has dominant and recessive genes. In addition to the loci there are also modifiers, which modify a certain gene. These include the rufus modifiers, color intensifiers, and plus/minus (blanket/spot) modifiers. A rabbit's coat only has two pigments, pheomelanin (yellow) and eumelanin (dark brown). There can also be no pigment, causing an albino or white rabbit.

Color Genes

Within each group, the genes are listed in order of dominance, with the most dominant gene first. In parenthesis after the description is at least one example of a color that displays this gene.

Chinchilla

- "A" represents the agouti locus (multiple bands of color on the hair shaft). The genes are:
- A= agouti ("wild color" or chestnut agouti, opal, chinchilla, etc.)
- a(t)= tan pattern (otter, tan, silver marten)
- a= self or non-agouti (black, chocolate)
- "B" represents the brown locus. The genes are:
- B= black (chestnut agouti, black otter, black)
- b= brown (chocolate agouti, chocolate otter, chocolate)

Blue

- "C" represents the color locus. The genes are:
- C= full color (black)
- c(ch3)= dark chinchilla, removes yellow pigmentation (chinchilla, silver marten)
- c(ch2)= medium (light) chinchilla, Slight reduction in eumelanin creating a more sepia tone in the fur rather than black.
- c(ch1)= light (pale) chinchilla (sable, sable point, smoke pearl, seal)
- c(h)= Himalayan, body white with extremities ("points") colored in black, blue, chocolate or lilac, pink eyes
- c= albino (ruby-eyed white or REW)
- "D" represents the dilution locus. This gene dilutes black to blue and chocolate to lilac.
- D= dense color (chestnut agouti, black, chocolate)
- d= diluted color (opal, blue or lilac)
- "E" represents the extension locus. It works with the 'A' and 'C' loci, and rufus modifiers. When it is recessive, it removes most black pigment. The genes are:
- E(d)= dominant black
- E(s)= steel (black removed from tips of fur, which then appear golden or silver)
- E= normal
- e(j)= Japanese brindling (harlequin), black and yellow pigment broken into patches over the body. In a broken color pattern this results in Tricolor.
- e= most black pigment removed (agouti becomes red or orange, self becomes tortoise)

Dutch

- "En" represents the plus/minus (blanket/spot) color locus. It is incompletely dominant and results in three possible color patterns:
- EnEn= "Charlie" or a lightly marked broken with color on ears, on nose and sparsely on body
- Enen= Broken rabbit with roughly even distribution of color and white
- enen= Solid color with no white areas
- "Du" represents the Dutch color pattern, (the front of the face, front part of the body, and rear paws are white, the rest of the rabbit has colored fur). The genes are:
- Du= absence of Dutch pattern
- du(d)= Dutch (dark)
- du(w)= Dutch (white)

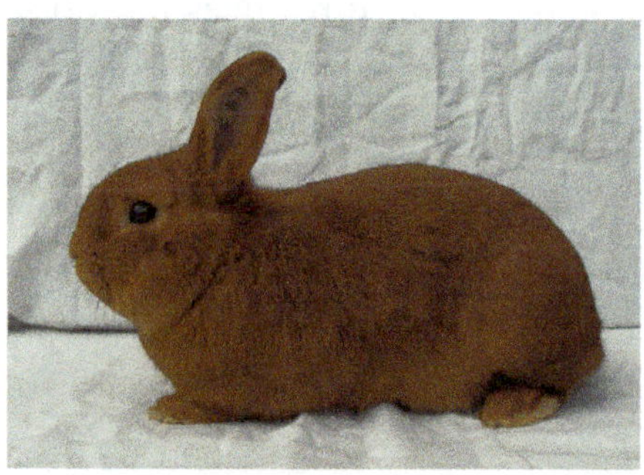

Czech Red

- "V" represents the vienna white locus. The genes are:
- V= normal color
- Vv= Vienna carrier, carries blue-eyed white gene. May appear as a solid color, with snips of white on nose and/or front paws, or Dutch marked.

- v= vienna white (blue-eyed white or BEW)
- "Si" represents the silver locus. The genes are:
- Si= normal color
- si= silver color (silver, silver fox)
- "W" represents the middle yellow-white band locus and works with the agouti gene. The genes are:
- W= normal width of yellow band
- w= doubles yellow band width (Otter becomes Tan, intensified red factors in Thrianta and Belgian Hare)
- "P" represents the OCA2 form of albinism, P is because it is an integral P protein mutation. The genes are:
- P= normal color
- p= albinism mutation, removes eumelanin and causes pink eyes. (Will change, for example, a Chestnut Agouti into a Shadow)

REFERENCES

1. Anthon, Charles. A System of Ancient and Mediæval Geography for the Use of Schools and Colleges pg.14. Books.google.com. Retrieved 2012-10-10.
2. Whitman, Bob D. (2004). Domestic Rabbits & Their Histories: Breeds of the World. Leawood, KS: Leathers Publishing. p. 450. ISBN 1-58597-275-4.
3. Dunlop, Robert H.; David J. Williams (1996). Veterinary Medicine: An Illustrated History. St Louis, MO: Mosby. p. 42. ISBN 0-8016-3209-9.
4. Whitman, Bob D. (2004). Domestic Rabbits & Their Histories: Breeds of the World. Leawood, KS: Leathers Publishing. pp. 342–343. ISBN 1-58597-275-4.
5. Whitman, Bob D. (2004). Domestic Rabbits & Their Histories: Breeds of the World. Leawood, KS: Leathers Publishing. pp. 341–344. ISBN 1-58597-275-4.
6. Whitman, Bob D. (2004). Domestic Rabbits & Their Histories: Breeds of the World. Leawood, KS: Leathers Publishing. p. 343. ISBN 1-58597-275-4.
7. Whitman, Bob D. (2004). Domestic Rabbits & Their Histories: Breeds of the World. Leawood, KS: Leathers Publishing. p. 346. ISBN 1-58597-275-4.
8. Whitman, Bob D. (2004). Domestic Rabbits & Their Histories: Breeds of the World. Leawood, KS: Leathers Publishing. pp. 347–350. ISBN 1-58597-275-4.
9. Bennett, Bob (2009). Storey's Guide to Raising Rabbits: Breeds, Care, Housing. North Adams, MA: Storey Publishing. pp. 45–49. ISBN 978-1-60342-456-1.
10. Whitman, Bob D. (2004). Domestic Rabbits & Their Histories: Breeds of the World. Leawood, KS: Leathers Publishing. p. 190. ISBN 1-58597-275-4.
11. Whitman, Bob D. (2004). Domestic Rabbits & Their Histories: Breeds of the World. Leawood, KS: Leathers Publishing. p. 15. ISBN 1-58597-275-4.
12. Druett, Joan. "Chapter Eight — Living with embarrassment: the rabbit". Exoctic

Invaders. New Zealand Electronic Text Collection. Retrieved 14 October 2012.

13. Whitman, Bob D. (2004). Domestic Rabbits & Their Histories: Breeds of the World. Leawood, KS: Leathers Publishing. p. 9. ISBN 1-58597-275-4.
14. Whitman, Bob D. (2004). Domestic Rabbits & Their Histories: Breeds of the World. Leawood, KS: Leathers Publishing. p. 68. ISBN 1-58597-275-4.
15. Whitman, Bob D. (2004). Domestic Rabbits & Their Histories: Breeds of the World. Leawood, KS: Leathers Publishing. pp. 48–49. ISBN 1-58597-275-4.
16. Anderson, Virginia DeJohn (2004). Creatures of Empire: How Domestic Animals Transformed Early America. New York: Oxford University Press. p. 97. ISBN 978-0-19-530446-6.
17. Whitman, Bob D. (2004). Domestic Rabbits & Their Histories: Breeds of the World. Leawood, KS: Leathers Publishing. pp. 151–152. ISBN 1-58597-275-4.
18. Whitman, Bob D. (2004). Domestic Rabbits & Their Histories: Breeds of the World. Leawood, KS: Leathers Publishing. p. 115. ISBN 1-58597-275-4.
19. Whitman, Bob D. (2004). Domestic Rabbits & Their Histories: Breeds of the World. Leawood, KS: Leathers Publishing. p. 139. ISBN 1-58597-275-4.
20. Dunlop, Robert H.; David J Williams (1996). Veterinary Medicine: An Illustrated History. St Louis, MO: Mosby. pp. 354–355. ISBN 0-8016-3209-9.
21. Whitman, bob D. (2004). Domestic Rabbits & Their Histories: Breeds of the World. Leawood, KS: Leather Publishing. p. 120. ISBN 1-58597-275-4.
22. Whitman, bob D. (2004). Domestic Rabbits & Their Histories: Breeds of the World. Leawood, KS: Leather Publishing. p. 88. ISBN 1-58597-275-4.
23. Whitman, bob D. (2004). Domestic Rabbits & Their Histories: Breeds of the World. Leawood, KS: Leather Publishing. p. 90. ISBN 1-58597-275-4.
24. Dunlop, Robert H.; David J Williams (1996). Veterinary Medicine: An Illustrated History. St Louis, MO: Mosby. p. 377. ISBN 0-8016-3209-9.
25. Whitman, bob D. (2004). Domestic Rabbits & Their Histories: Breeds of the World. Leawood, KS: Leather Publishing. p. 376. ISBN 1-58597-275-4.
26. Templeton, George S. (1968). Domestic Rabbit Production. Danville, Illinois: The Interstate Printers & Publishers. pp. 61–63.
27. Whitman, Bob D. (2004). Domestic Rabbits & Their Histories: Breeds of the World. Leawood, KS: Leathers Publishing. p. 228. ISBN 1-58597-275-4.
28. Whitman, Bob D. (2004). Domestic Rabbits & Their Histories: Breeds of the World. Leawood, KS: Leathers Publishing. pp. 441–443. ISBN 1-58597-275-4.
29. Whitman, Bob D. (2004). Domestic Rabbits & Their Histories: Breeds of the World. Leawood, KS: Leathers Publishing. pp. 425–429. ISBN 1-58597-275-4.
30. Hreiz, Jay (May–June 2012). "Dr.". Domestic Rabbits. 40 (3): 75.
31. Whitman, Bob D. (2004). Domestic Rabbits & Their Histories: Breeds of the World. Leawood, KS: Leathers Publishing. p. 448. ISBN 1-58597-275-4.

32. Templeton, George S. (1968). Domestic Rabbit Production. Danville, Illinois: The Interstate Printers & Publishers. p. 17.
33. Fish & Wildlife Service. "Press Release 14 Jan 1943" (PDF). Retrieved 15 October 2012.
34. Ashbrook, Frank G. (1943). How To Raise Rabbits for Food and Fur. New York: Orange Judd. pp. 23–28.
35. T., A. (1 April 1944). "Some Remarks on the History of the Rabbit in Australia". The Argus. Retrieved 15 October 2012.
36. Whitman, Bob D. (2004). Domestic Rabbits & Their Histories: Breeds of the World. Leawood, KS: Leathers Publishing. p. 119. ISBN 1-58597-275-4.
37. Beeman, Joseph. "Site of U.S. Rabbit Experimental Station". Historical Marker Database. Retrieved 15 October 2012.
38. Lebas, F. (1997). "9". The Rabbit: Husbandry, Health and Production. Rome: Food Agriculture Organization.
39. Foster, M (September 1996). "Structure of the Australian Rabbit Industry" (PDF). ABARE Report. pp. 8–9. Retrieved 15 October 2012.
40. Lebas, Francois. "Constitution of the World Rabbit Science Association". World Rabbit Science Association. Retrieved 14 October 2012.
41. Abbott, Allison (2 August 2012). "Court Orders Temporary Closure of Dog-Breeding Facility". Nature. doi:10.1038/nature.2012.11121.
42. "HRS Activist Corner". House Rabbit Society. Retrieved 15 October 2012.
43. McCarthy, Bob (11 August 2011). "Anonymous Call to New Animal Abuse Hotline Leads to Raid on Colorado Woman's Rabbit Farm". Breithart.com. Retrieved 15 October 2012.
44. Hottle, Molly. "23 Rabbits Stolen from Portland Meat Collective Farmer". The Oregonian. Retrieved 15 October 2012.
45. Olivia Geng, French Rabbit Heads: The Newest Delicacy in Chinese Cuisine. The Wall Street Journal Blog, 2014-06-13. (The original text has 200 million tons for the world production, but this has to be a typo for 2 million, in order for the China numbers given in the same article too match, as well as to be in line with other sources.)
46. The rabbit - husbandry, health and production
47. "How to Cook Everything :: Braised Rabbit with Olives". 2008. Archived from the original on 2008-05-17. Retrieved 2008-07-17.
48. Kulpa-Eddy. "A review of trends in animal use in the United States (1972–2006)" (PDF). AATEX. Proc. 6th World Congress on Alternatives & Animal Use in the Life Sciences (14, Special Issue): 163–165.
49. Morton, Daniel (April 1988). "The use of rabbits in male reproductive toxicology". Environmental Health Perspectives. National Institute of Health. 77: 5–9. doi:10.2307/3430622. JSTOR 3430622. PMC 1474531 . PMID 3383822.

50. M.K. Prinsen (2006). "The Draize Eye Test and in vitro alternatives; a left-handed marriage?". Toxicology in Vitro. 20 (1): Pages 78–81. doi:10.1016/j.tiv.2005.06.030. PMID 16055303.

51. "American Livestock Breeds Conservancy: Belgian Hare". Albc-usa.org. Retrieved 2012-10-10.

52. Hare Surveyhttp://www.belgianhareclub.com/hare_survey.html

53. "Official Show Rules". American Rabbit Breeders Association. 2009.

Index

S

T

V